Foreword

THE OPERATIONS EVALUATION DEPARTMENT (OED) is an independent unit established within the World Bank to conduct reviews, on a selective basis, of Bank-supported programs, projects, and operational policies.[1] Such reviews attempt to determine whether the objectives of these programs and policies are being realized and how they might be made more effective, efficient, and responsive to the needs and concerns of member countries. The department reports directly to the Bank's Board of Executive Directors, and its reports are made available to all member countries of the Bank.

The role of lending by the World Bank is not simply to transfer resources, but also to render assistance in solving problems associated with development in member countries. Therefore, the evaluation undertaken by OED has to be concerned with activities at different levels of both the Bank and its member countries and has to be responsive to the concerns of each.

While preserving their professional independence, OED staff endeavor to maintain close contacts with relevant Bank staff and government officials so that their views are adequately considered in OED reports, so that the analysis and findings in these reports are clearly communicated, and so that OED is kept informed of responses to its findings in the Bank and in member countries.

This study is being published in the belief that it will add to the wealth of information provided in other reports published recently by the World Bank on this subject. This study, it is hoped, will also contribute to the promotion of creative approaches in this field.

The study was conducted by an international team of ten consultants drawn from universities and national and international research organizations. The team was led by Salem Gafsi of the Operations Evaluation Department and consisted of Eric A. Abbott (University of Iowa), George Alcorn (Extension Service in California), Robert F. Chandler (International Rice Research

1. The term World Bank includes the International Bank for Reconstruction and Development (IBRD) and the International Development Association (IDA).

Institute), Ralph Cummings, Sr. (Consultative Group on International Agricultural Research), John Fett (University of Wisconsin), Peter Oram (International Food Policy Research Institute), Subhi Qasem (University of Jordan), Guy Ronanet (Institut de Recherche Agronomique et Tropicale, France), Romesh S. Sood (Ministry of Agriculture, India), and Burt Swanson (University of Illinois).

The cooperation rendered by individuals and agencies in the ten countries of case study and the assistance provided by the Bank staff in the preparation of this report are greatly appreciated.

SHIV S. KAPUR
Director-General
Operations Evaluation

8.05

Presented by R Edwin Shanks

Agricultural Research and Extension

An Evaluation of The World Bank's Experience

THE WORLD BANK
Washington, D.C., U.S.A.

First printing June 1985

Manufactured in the United States of America

Library of Congress Cataloging in Publication Data

World Bank. Operations Evaluation Dept.
Agricultural research and extension.

1. Agriculture—Research—Developing countries—Evaluation. 2. Agricultural extension work—Developing countries—Evaluation. 3. Agricultural development projects—Developing countries—Evaluation. 4. Agriculture—Developing countries—Technology transfer—Evaluation. I. Title.

S542.3.W67 1985 630'.720172'4 85-3322
ISBN 0-8213-0510-7

Contents

Preface

THE CONCLUSIONS IN THIS STUDY are based on a review of the World Bank's support for agricultural research and extension (R&E) in 128 projects in ten member countries over the period 1974–80.

The projects reviewed include agricultural projects with research or extension components or both, and freestanding research or extension projects. At the time of the study most of the former had been completed and the majority of the latter were beyond the midpoint of their implementation period. The 128 projects reviewed represented 35 percent of the total number of projects that received the Bank's support for research and extension; the freestanding national research or extension projects reviewed accounted for 55 percent of the projects supported by the Bank.

The ten countries included in the review were Brazil, India, Indonesia, Kenya, Mali, Morocco, Nigeria, Sudan, Thailand, and Turkey. The countries were selected to permit coverage of (1) the various approaches the Bank has followed in providing support for research and extension; (2) all geographical regions as defined by the Bank for internal administration; (3) major agroecological zones; and (4) countries in which the Bank has supported research and extension for a period long enough to permit an analysis of the experience.

The review examines the national organizations in research and extension and attempts to establish how well the Bank's assistance has benefited the selected member countries' overall efforts at building up their research and extension capabilities. This study identifies those characteristics of countries that enhanced or hindered the effectiveness of the Bank's support for research and extension. These include the countries' resource endowments (physical, financial, and institutional) and the policies pursued in the agricultural sector.

The decision to undertake this review resulted from concern within the institution over the low level of support from the Bank for research and extension and over the relative efficiency of various approaches the Bank has adopted in providing such support. In addition, there was concern over the continued inadequacy of the services provided by the national research and extension organizations in support of Bank-financed agricultural and rural development projects and over the limited impact the Bank's support for research or extension has had in countries with weak national institutions.

Most of the Bank's lending for research and extension in the 1960s and early 1970s was done within the framework of integrated agricultural development projects in which attempts were made to strengthen R&E services together with input supply, marketing, credit, and rural infrastructure. The effectiveness of this approach was questioned in the early 1970s, when as a result of lessons derived from past experience, the Bank adopted its poverty alleviation policy in rural development and made some important shifts in its agricultural lending programs. Among these important changes was a substantial increase in the support it provided to freestanding research and extension projects in those countries where such support was considered appropriate. The aim of these projects was to bring effective and relevant low-cost technologies to small producers.

Over the past eight to ten years, the Bank has allocated substantial financial and institutional resources to support the promotion and dissemination of the training and visit (T&V) technique of extension. Because this study is primarily concerned with the research and extension functions within the overall effort for agricultural and rural development, the research methodologies and extension systems (including T&V) addressed here are considered solely from the point of view of how a particular research methodology or extension technique fitted overall with the approach the Bank had selected for countries with varying socioeconomic and institutional environments and differing needs and priorities.

Experience with the T&V technique of extension, as with others, suggests that at least ten to fifteen years are needed to get the system firmly in place and to develop a professional field-based system. An overall and substantive evaluation of the impact of T&V, therefore, has to await the completion of a number of projects the Bank has financed in support of this extension technique. In the meantime, the Bank has co-sponsored with the United Nations

Development Programme and the Food and Agriculture Organization two workshops at which users of T&V in Asia exchanged their experiences. The proceedings of these workshops, held in Thailand and Indonesia, have been published by the Bank.[1]

Another area of special concern to the Bank has been the research and extension needs of sub-Saharan Africa. Some agricultural research of good quality has been done in this part of the world, mainly on a commodity basis. But scarcity of funds and of trained manpower has limited the application of research results. The particular problems of sub-Saharan Africa have been addressed in this study through a review of Bank-supported programs in Kenya, Mali, and Nigeria. Subsequent to this study, the Bank in cooperation with member countries and other international organizations has undertaken several special reviews of research and extension programs aimed at gaining better understanding of existing systems in sub-Saharan Africa to prepare more effective research and extension projects.

1. Michael M. Cernea, John K. Coulter, and John F. A. Russell, eds., *Agricultural Extension by Training and Visit: The Asian Experience* (Washington, D. C.: The World Bank, 1984), and *Research-Extension-Farmer: A Two-Way Continuum for Agricultural Development* (forthcoming).

Agricultural Research and Extension

An Evaluation of The World Bank's Experience

Major Findings and Recommendations

THE WORLD BANK'S EFFORTS in support of research and extension (R&E) were significant and worthwhile, even though the efficacy of the approaches followed in various countries to address specific needs differed. This study proceeded from the premise that research and extension were related functions and examined the Bank's initiatives in this light. The study found that the Bank's support for R&E, which is still evolving, has helped highlight the importance of strengthening these two functions.

The Bank's experience with R&E has been positive to date. An important feature of the Bank's support over the last ten years has been the sustained efforts the Bank's staff has made to strengthen the role of national organizations in research and extension. Because circumstances were difficult, it might have been easier, in many cases, to wait for conditions to become favorable before extending such support. Instead, the Bank endeavored to help create the necessary favorable conditions. This step is commendable. The lessons learned so far about the strengths and weaknesses in the approaches selected for supporting R&E have uniquely equipped the Bank to be effective in helping member countries improve the efficiency of their R&E organizations.

Experience in the ten member countries examined for this study showed that the Bank tried three approaches to help create better conditions for increasing productivity and production in agriculture through agricultural research and extension.

The first approach consisted of supporting research or extension components of agricultural and rural development projects. These components were designed and implemented independently of the national R&E organizations. The primary purpose of these components was to back up production-oriented activities in the project area within the project investment period. Project documents usually stipulated that at the end of the investment period the national R&E organizations would be responsible for continuing to provide R&E services, though not necessarily in the same format or intensity as under the project.

The second approach was similar to the first with the exception that the national R&E organizations were partly or fully associated in implementing the component, but only rarely associated in designing it.

The third approach consisted of providing support to full-scale research and extension projects and to the national R&E organization. Under this approach, the emphasis placed was both on the immediate support for agricultural production during the investment period and on longer-term institution building during and beyond the initial investment period. Although the length of the investment period of full-scale R&E projects was similar to that of projects with R&E components, the Bank recognized that benefits from full-scale projects would take longer to materialize. For this reason, many of these projects were designed to be just one phase in the longer-term process of building a national R&E capability.

The Bank's support for R&E shifted from components in agricultural projects to freestanding national projects in those countries in which policies and institutions were found favorable to strengthening R&E functions. The Bank's *Agricultural Research* sector policy paper of 1981 recommended improvements in the design of R&E components, their phasing over time, and their linkages to national R&E systems. The paper also suggested that the broad-based support given to R&E through national projects be improved to make them an effective tool to meet countries' technology needs.

This study has found merit in the Bank's flexible policy, which has allowed its staff to tailor the approach to R&E to individual countries' diverse needs. The study team exhorts the Bank to continue its support for both components and freestanding projects in research and extension.

The review found that, regardless of the approach followed in the ten countries studied, the level of R&E services needed to achieve the countries' development targets far exceeded the level of services provided. This gap was a result of the fact that the countries' development objectives were not fully related to the means available for achieving them and of the fact that R&E services were generally provided in a fragmented manner.

The Bank's efforts to support the development of R&E functions were often hindered by the limited response it received from policymakers with regard to formulating and implementing the necessary actions. This limited response had two origins. First, the limited capability in some countries to effectively identify, prepare, and appraise R&E components and projects meant that they did not fully appreciate the urgent need for policy actions. Second, the Bank's suggestions for policy actions in support of R&E sometimes failed to reflect realistically existing political, institutional, and financial constraints in the countries.

The study found inadequacies in resource allocation to and between research and extension; this reflected weaknesses in planning and monitoring processes in the countries. Frequently, development plans and project documents showed more concern with the quantity of resources allocated for R&E than with the effectiveness and impact of their use. The Bank attempted to help countries make better use of R&E resources in several ways: through its broad-based direct support to R&E systems, through its cooperation with other outside agencies financing activities in support of R&E, and through its policy dialogue with members. In some countries these efforts were successful; in others they were less so. Broad-based support for R&E has created additional demands for scarce budgetary resources, which some countries are unable to sustain. On the whole, the experience with cofinancing for R&E has been positive, although the amounts involved have been limited. In the policy dialogue, the Bank has effectively used R&E components in agricultural and rural development projects in several countries as the precursor to freestanding R&E projects.

In *research*, the countries studied felt that the Bank had a positive impact on resource allocation to and within research systems. However, the Bank did not always consider the ability of the country to sustain improvements made in research services beyond the investment period of the projects it supported. While the Bank has successfully supported the development of physical research facilities, this success has not yet been matched by improvements in the management of these facilities or by the development of institutional arrangements that would be conducive to their proper use.

In *institutional development*, the Bank's efforts have been hindered by the countries' limited resources and unclear priorities, which have made policymakers reluctant to support proposals that have less visible benefits in the long term and more visible costs in the short term. Considering this reality, the Bank's support for research components in agricultural projects has been worthwhile, particularly because it has kept the dialogue going on how to strengthen research in the countries.

In *applied research*, although the Bank's concentration was usually justified, officials in many countries thought the Bank should broaden its support to help make research more responsive to sectorwide needs. These officials wanted a broader coverage of crop and livestock activities, support for basic research through national, regional, or international arrangements, and help in improving the efficiency of adaptive research services. The Bank does provide substantial support for basic and applied research on a wide variety of crops and livestock at various international research centers through the Consultative Group on International Agricultural Research.

Because some countries are unable to monitor the efficiency and usefulness of their research programs, the Bank promoted the establishment of monitoring systems. The Bank has yet to transform its own project supervision system into a monitoring process, however. Poor monitoring of research has led to duplication of research activities within countries and between countries and international research centers; it has also resulted in a lack of accountability for, and concern about, the use of research results.

In *extension components*, the Bank's support has been worthwhile. The support has helped meet the urgent and realistic needs of developing member countries and has increased their awareness of the potential benefits of technology transfer. These extension components, however, have tended to stress the technology supply and service aspects of the extension function, often at the expense of its education aspect. This emphasis has been a result of limited correlation between the scope and substance of extension components and the state of the research function in the country; it is also a reflection of limited emphasis on proper resource management.

The major strength of Bank-supported extension through agricultural projects was its ability to increase the access of groups of farmers, particularly small farmers, to an integrated package of technology. The package, which included physical inputs, credit, technical advice, and other services, proved especially useful for farmers involved in the production of a variety of crops and livestock. Technology could be transferred to these farmers under conditions of limited or unskilled manpower and weak national research. During the implementation of many extension components, however, the provision of inputs and services tended to prevail over the actual extension either because there was no need for actual extension, owing to the simplicity of the technology used and the farmers' prior familiarity with it, or, where extension really was needed, because of the lack of suitably trained extension personnel.

The full-scale extension projects supported by the Bank covered a wider area and a larger number of farmers than did the extension components in agricultural and rural development projects. The full-scale projects sought to introduce organizational change and to improve the performance of extension workers. This approach appeared to suit countries with large numbers of small farmers at a low level of technological development and with abundant but poorly trained and inadequately used extension staff. These projects, however, often made unrealistic assumptions about the availability and quality of technical and extension specialists. As a result, limited effort was made to significantly upgrade the competence of specialized personnel and to expand their availability.

Where the Bank supported full-scale extension projects, it sought to have the extension function separated from that of input supply and other services so that extension workers could concentrate on delivering a technical message without being distracted by administrative, supply, or regulatory duties. Because some of the countries lacked an alternative private-sector mechanism for providing essential inputs and services, however, the redefinition of the extension function left some farmers less well served. In addition, the concentration of efforts on delivering technical messages appeared, at times, premature because of the dearth of suitable technical messages flowing from research. Under such circumstances, an integration of the technology transfer functions appeared to have worked well as a first phase in the evolution of extension services. An integrated approach to the supply of technical advice, physical inputs, and credit through the extension worker can work. Changes in the extension organization are not always necessary to improve the efficiency of an existing technology transfer system.

The study found that the design of Bank-supported full-scale extension projects did not always assess the suitability of the extension methodology in the country's social and cultural environment. Nor did the design consider whether trained personnel and financial resources were sufficient to sustain the expanded extension service beyond the life of the project. There often were inconsistencies between the time period within which the Bank's support was provided and the time required for the extension programs to achieve the objectives set for them. There is now a clear recognition that reorganization and upgrading of extension services covering large areas and involving large numbers of personnel constitute, inevitably, a time-consuming effort that requires sustained support.

The review found that regardless of the approach followed, Bank support for R&E has been constrained by five factors: (1) lack of clarity in the country's objectives for agriculture and in the priorities among them; (2) limited input by the country in the design of R&E components and projects, along with unclear links between R&E activities supported by the Bank and other development activities in the sector; (3) limited sectorwide or economy-wide work on issues affecting R&E; (4) institutional separation of research and extension; and (5) lack of clarity in, or agreement on, the definition of various stages in the process of technology development and transfer. The review found that, depending on the approach selected for providing support to R&E, the Bank has succeeded to some extent in making policymakers aware of the need to deal with these constraints and in helping to resolve some of them. A brief comment on each of these factors follows.

First, the development aspirations of several countries were beyond the means and the immediate potential for development. In addition, weak planning and varying degrees of political instability left development objectives unclear and development priorities undefined.

Second, the capacity for local participation in the preparation and evaluation of projects (at both the policy level and the technical level) was weak in several countries; this resulted in frequent and substantial input by the Bank in the design of R&E

components and projects. Although the Bank's effort was appreciated by policymakers and planners, it was not always welcomed by the technical staff in the implementing agencies.

Third, there were weaknesses in the coordination of support for R&E by the Bank with the support for related improvements in supply and distribution of inputs, credit, marketing, training, and so forth (from the Bank or elsewhere). The Bank's support for R&E often strained countries' human and financial resources because the project designs did not (and probably could not) adequately take into account the other long-term demands on these resources.

Fourth, the technology development and transfer functions were, in many cases, treated in isolation from each other by both the countries concerned and the Bank. This constraint hindered the strengthening of the process of technology development and transfer. Organizations responsible for related functions were often institutionally separated. Psychological barriers frequently developed among those organizations because their functions overlapped or their terms of reference were undefined and because they had to compete for scarce budgetary resources.

Fifth, the Bank limited its support to research activities at the adaptive end of the research continuum, even in countries that needed improvements in research of a more basic nature.[1] Similarly, extension projects or extension components often did not include a critical assessment of the technology to be extended (particularly of its complexity relative to the existing technology) or of the technology transfer function (education, communication, technology supply, and technology service).

The more successful research organizations in the ten countries studied displayed a combination of some of the following characteristics:

- A strong central organization responsible for most aspects of national research and backed up by sufficient funds
- Some degree of autonomy from the routine bureaucratic structure of the Ministry of Agriculture

1. Some Bank staff feel that the Bank should not get involved in basic research that is most appropriately carried out at universities. In the projects supported, the Bank has stressed the problem-solving character of research.

- Good links to the national planners and policymakers despite the semiautonomous or autonomous status
- Service units—either directly part of, or closely related to, the central research organization—which perform important planning, monitoring, and evaluation functions
- Planned decentralization of regional research
- Active participation in translating research results into recommendations for users.

The more successful extension organizations displayed a combination of some of the following characteristics:

- Flexible extension methodology to fit the nature of the technology being extended, clients' needs, and the means available to the country
- Close coordination of the extension function with other technology transfer functions
- Active participation by extension personnel (subject matter specialists), along with researchers, in translating research results into recommendations for users
- A formal feedback mechanism that directly reflects clients' points of view rather than observers' opinions of what those points of view are or should be (this may require training clients to work together in articulating their demands more clearly to research and extension staff)
- Clearly defined job descriptions for extension personnel, who are supervised and evaluated on the basis of these job specifications
- Adequate attention to both the technical skills and the communication or pedagogical skills needed by extension personnel in view of the technology to be extended and the clients to be served.

In those developing countries that have limited capability for project preparation, the Bank could make its support for R&E more relevant by:

- More in-depth assessment of the social, political, and cultural constraints in the country, to make the projects designed more consistent with the realities in the country.

- More insistence on borrowers' active participation in the preparation and appraisal of R&E projects or components. Although some countries may lack technical and management expertise, local staff are, nevertheless, in a better position than outsiders (consultants or Bank staff) to judge whether an R&E strategy or methodology fits their culture and environment.
- Less emphasis on uniform approaches to R&E or specific R&E techniques. Although the Bank's staff may understand such concepts as applied research, target groups, appropriate technology, training and visit, and appropriate organization, R&E administrators and consultants to governments, in their eagerness to identify "acceptable" projects for the Bank's support, may interpret these concepts in ways that may not be useful to the country. In discussing the use of uniform approaches, the Bank's staff should consider how the various concepts involved will be interpreted and used in existing R&E services and how the changes will be made if needed.

The design of R&E projects should start with a clear definition of the scope of the functions to be supported and an assessment of how they would relate to other functions in the development and transfer of technology. Such an assessment could be a basis for discussions between the Bank and the country on actions that could be undertaken by the country outside of the project to complement and reinforce the project.

An assessment of a country's ongoing R&E efforts and its goals in the agricultural sector should precede any support by the Bank for R&E in that country. The assessment should define the link between the Bank's lending for R&E and its lending for other agricultural development activities, as well as the link between Bank-supported operations in agriculture and the rest of the agricultural development program in the country. The Bank's economic and sector work is the obvious vehicle for attempting to integrate support for R&E in the Bank's agricultural lending program and to coordinate Bank support for R&E with other sources of support.

In supporting technology transfer, the Bank should consider how a balance could be achieved between the extension function and that of input supply and services that would best meet the needs of the population to be served. Whether all technology transfer functions should be located spatially and organizationally together or be separated will depend on specific circumstances.

In providing support for extension, the Bank should ensure that the projects are designed in a way that reflects the state of research in the country and that they take maximum advantage of the work of international research centers. The projects should also reflect the heterogeneity in the level of development of various sections of the agricultural sector and should be designed with enough flexibility to meet the varied demands of such sections.

Research projects should emphasize:

- Manpower training, and the training of research program leaders in particular
- Development of the functions of research management and research station management
- Research on agricultural activities in low-potential areas
- Research that helps integrate crop production and livestock activities and improves farm management
- Consideration of social, economic, environmental, and political factors in the design of research programs and the evaluation of their results.

Extension projects should emphasize:

- Formal training for higher-level extension personnel in technical agricultural subjects as well as in communications
- Job descriptions for extension personnel that are consistent with the country's administrative procedures and the existing reward system
- Training that relates to (1) the country's demand for skilled manpower, its training capacity, and its budgetary limitations both during and after the project's completion, (2) the differences in the levels of technology being used by various groups of farmers, and (3) the potential for tradeoffs between the use of human resources and the communication media for extension
- Formal professional links between research and extension

- The establishment of channels through which farmers, especially the less privileged among them, can influence the form and substance of extension being provided.

The Bank's supervision of R&E projects could be improved substantially by providing more technical expertise to monitor the implementation of R&E components and projects. The Bank's supervision effort, however, should go beyond the physical implementation of the project; it should cover the quality of research programs and the institutional changes being introduced. Organizational changes, in particular, should be closely monitored and their usefulness periodically reassessed to ensure that the agencies affected continue to work in harmony with other national institutions. These measures, of course, will affect the supervision coefficients and the profile of the staff in charge of the supervision function.

1

Objectives and Methodology

THE SUPPORT PROVIDED BY THE WORLD BANK for agricultural research and for the delivery of its results to farmers has been increasing since the mid–1960s. This support has taken several forms: full-scale projects in research or extension or both; research or extension components in agricultural and rural development projects; agricultural university components in education projects; and financial and institutional support for international centers for agricultural research through the Consultative Group on International Agricultural Research (CGIAR).

Between fiscal 1970 and 1980, the Bank committed more than $745 million in loans and credits in support of twenty-seven full-scale research or extension projects in thirteen countries.[1] In addition, financial support amounting to $368 million was provided over the period fiscal 1970–80 for research or extension components in 312 agricultural and rural development projects in eighty countries. Commitments of $51 million over the same period were also made for research components financed as part of eighteen education projects in twelve countries. Finally, support of close to $55 million was provided to the CGIAR during 1972–80. This puts the total funds made available by the Bank for research and extension over the past decade at $1,219 million.

The scope and intent of the Bank's support for R&E projects can be classified in four categories. In the first category, the Bank sought to strengthen already established national R&E services by providing them with the means to increase their capability to conduct relevant research and to transfer the results to farmers. The Bank expected that its support would make a measurable impact on agricultural production during the period of the funding and beyond. In the second category, the Bank sought to help existing but weak national R&E services get established; support was

1. The fiscal year of the World Bank runs from July 1 to June 30. All money terms are expressed in U.S. dollars.

provided for their institutional development and for constructing R&E facilities. The immediate result expected was the establishment of more capable agencies that could generate results beyond the period of the Bank's funding. In the third category, the Bank provided support for R&E specifically to help achieve agricultural production targets set in investment programs. Such support was provided either outside the national R&E system or at its margin. The funds allotted for R&E in this category were expected to have an impact on production during the period the investment program was being implemented; there was usually little concern for longer-term needs. In the fourth category, the Bank sought to combine short-term and long-term objectives of an institution-building and production-increasing nature.

Conflicting opinions have been expressed on the merits of these approaches in supporting R&E. Both inside and outside the Bank, questions have been raised about the effectiveness of the funds provided by the Bank in support of R&E. The slow progress made in increasing agricultural productivity, coupled with the perception that in the past funds provided for R&E tended to be below requirements, has led to the widely held conviction that increased support for R&E (from various sources) would have to be a major part of any program aimed at increasing productivity in agriculture in the 1980s.[2]

Objectives and Guidelines

In response to the concerns and questions raised in the preceding section, the Operations Evaluation Department decided to undertake a review of the Bank's past experience in supporting R&E. The aim was to identify strengths and weaknesses in various approaches the Bank had selected for supporting R&E systems and to analyze the underlying reasons for such strengths and weaknesses. An effort was made to look at the Bank's approach rather than at the results of individual R&E projects. It was hoped that such an analysis would be a basis for formulating recommendations

2. See, for example, World Bank, *Agricultural Research*, Sector Policy Paper (Washington, D.C., June 1981).

specific to each member country to enhance the effectiveness of its R&E and for developing guidelines for future support for R&E by the Bank.

The study was planned around four major topics:

- How did the Bank's support for R&E relate to development goals and programs in various countries, to their research objectives and programs, and to the Bank's own lending strategies and programs?
- What is the logic behind the regional and commodity allocation of Bank funds earmarked for supporting R&E systems? The geographical distribution of Bank support for R&E has sometimes been perceived as not adequately reflecting the relative urgency of the needs of various R&E systems and their access to alternative sources of support, while the commodity distribution has been perceived as being focused too narrowly on a few commodities.
- Do the Bank's present procedures permit adequate planning of its support for R&E to fit the particular circumstances of a country and to allow for proper coordination with other financing agencies? The Bank has been concerned that its support for R&E complement the support provided for the same purpose by its borrowing member countries and by other outside financial agencies. After all, research and extension are parts of the same continuum; therefore, the timing and sequencing of support to each service determine the effectiveness of the total support for R&E regardless of its source. Moreover, because the Bank does not have sufficient resources to provide comprehensive support to R&E in every member country, the quality of its support is heavily affected by whether or not support from other agencies materializes and by the quality of such support.
- How has the Bank's approach to supporting R&E affected the establishment of links among R&E, the R&E system, its clients, and other institutions providing complementary services in the development process? The primary purpose of research is to develop technology for use in a given environment, while the primary purpose of extension is to transfer such a

technology to users. There is, therefore, a need for coordination of these two functions. The effectiveness of the Bank's support for these two functions is contingent upon (1) the existence of recognized and operating channels of communication among the institutional elements of the R&E system and between the system and its clients, and (2) the quality of the information flowing from one element of the system to another and the capability of other elements to use such information to make their services more responsive to the needs of their clients.

Methodology

The study was based on a review of the Bank's support for R&E in ten countries: Brazil, India, Indonesia, Kenya, Mali, Morocco, Nigeria, Sudan, Thailand, and Turkey. In selecting these countries, an attempt was made to ensure that each geographical region was represented in the sample; that major ecological zones were covered; that countries receiving substantial support from the Bank in their economic development and in strengthening R&E were included; and that the various approaches the Bank followed in providing support for R&E were represented.

The review was undertaken in two stages. First, a desk review was made of project appraisal and supervision reports (of completed and ongoing projects), project performance audit reports, agricultural sector studies, country agricultural development plans, and other relevant documents in the countries involved. Second, field visits were undertaken in the ten countries to discuss issues identified in the desk study and to seek reactions to these issues from people involved in the implementation of R&E projects in these countries.

The most revealing criterion for evaluating the efficiency of an R&E system is its output, which is knowledge and technology that meet the needs of the clients. This criterion could not be applied in this review for two reasons. First, in contrast with well-established systems, Bank-supported R&E systems were often at an early stage of development or in the process of undergoing substantial reform to reorient them to the development needs of their respective countries. In both cases, the output expected from the Bank's support for R&E was primarily an increase in institutional strength—that is, a buildup in the capability of R&E systems to develop and transfer technologies appropriate for their clients. Second, most Bank-supported R&E programs were in the process of being implemented at the time of the study. Therefore, the quantitative indicators of productivity were difficult to estimate.

Moreover, even when these indicators were available, an assessment based only on such indicators could be misleading at this stage for at least two reasons. First, a system should be assessed on the basis of its potential to contribute on a continuous basis once temporary and internally nonsustainable factors, such as an infusion of funds and technical assistance from the outside, have disappeared. Second, there is a time lag of up to ten years between the moment usable research results materialize and the moment their full impact on production is realized.

In addition, because this study is concerned only with an evaluation of the Bank's support for R&E, there is a conceptual problem in separating the impact of support from the Bank from the impact of support from other institutions. It is, therefore, difficult to establish a causal relationship between indicators of performance and the extent of support from the Bank.

For these reasons, it was decided to rely on those indicators of the Bank's support that could be used to assess the adequacy and efficiency of the Bank's approaches in meeting the specific needs of R&E in member countries. The indicators used throughout this report relate to three characteristics that were found to be critical to the efficiency and productivity of any R&E system.

- On the research side, the crucial resource is research scientists. The scientists' efficiency depends on the extent to which they are (1) informed of the problems facing the extension personnel, the farming community, and the country, (2) qualified and equipped to establish priorities and to develop plans to seek practical and appropriate solutions to research problems, and (3) professionally and materially motivated to perform their development-oriented functions.
- On the extension side, the main elements consist of qualified extension personnel and established channels and procedures for (1) transferring proven technology to potential users,

(2) monitoring the acceptance of that technology, and (3) feeding back into research the relevant information collected from the monitoring process.

- Concerning both research and extension, it is essential that the two functions be recognized as tools for development. Therefore, the R&E system should (1) have defined targets based on needs agreed upon by clients and policymakers, (2) be allocated the means to achieve those targets, and (3) be held accountable for ensuring that the results are commensurate with the resources allocated.

Framework and Definitions

The system for the development and transfer of technology refers to those elements of national institutions responsible for research or extension or both; it also refers to the internal linkages between those elements and the external linkages to national and international organizations. While recognizing that in practice research and extension generally are organized separately, a comprehensive review of the two functions and their linkages requires that all aspects of the development and transfer of technology be considered as stages in a continuum. The development and transfer of technology belong to a two-way process that could effectively link policymakers, researchers, extension agents, and users of agricultural technology in one integral system (see chart). The following definitions attempt to clarify how various important terms are used in this book.

Agricultural technology includes both the *components* and the *processes* of agricultural production. Components include such items as seeds, fertilizer, pesticides, and machinery. Processes include the elements needed for the optimum combination of those components by the producer; these include information about components and the technical knowledge and skills to use them effectively, and management skills and the know-how necessary to adapt the use of the components to a given production environment or to given production goals.

Agricultural research, a function that falls within the purview of national technology development, consists of the scientific process

The Technology Development and Transfer System

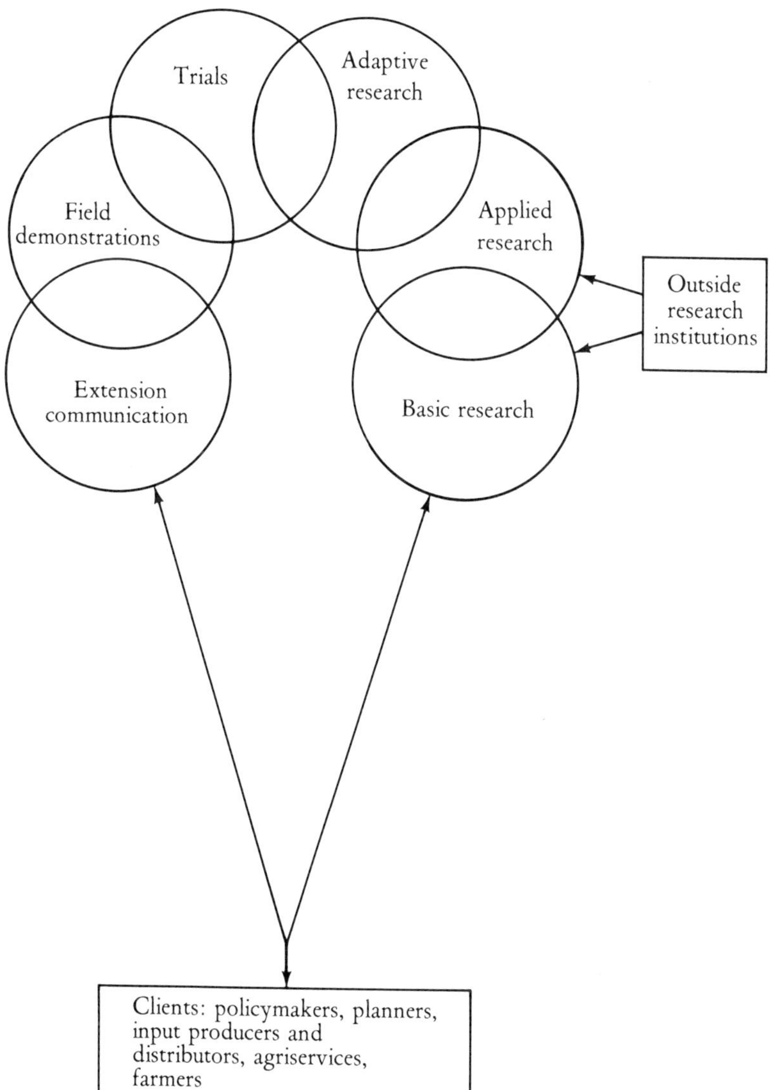

of generating new knowledge and translating new and existing knowledge into new technology. Agricultural research is performed at different points along the technology development continuum. The exact nature of the research depends on the need and the means (including qualified research scientists) available. Research is commonly classified as basic, applied, or adaptive. *Basic* agricultural research is viewed as the process of generating new knowledge. It attempts to explain why certain observed phenomena happen. Examples include research on insect physiology and ecology, to further understanding of the life cycle of insects, and research on the basic physiological and biological functions of plants and animals. *Applied* agricultural research is viewed as the process of finding practical use for existing knowledge. It attempts to identify ways to take advantage of known scientific facts. Examples include breeding plants and animals for obtaining higher yields or for developing resistance to certain diseases and research on soil fertility. *Adaptive* agricultural research is viewed as the process of tailoring the broadly based relevant findings from applied research to the requirements of specific locations. Adaptive research can be performed both at research stations and on selected farms. On-farm adaptive research should not be confused with on-farm trials and on-farm demonstrations, which aim at testing technology and showing potential users the merits or weaknesses of different practices. On-farm adaptive research and trials do have a demonstration effect; however, this is a by-product of testing a technology in different locations.

The *technology transfer* service performs such functions as informing producers of the existence of improved technological components and processes, making these available to potential users, and assisting them in learning how to take advantage of improved production practices. Technology transfer is a process with multiple functions; these include *information, teaching, technology supply*, and *technology service*. These functions can be performed by the same organization, or by separate organizations, as long as their complementarity is preserved, their separate objectives are understood, and the organizations have the capacity and the means of performing them.

Agricultural extension is primarily concerned with the information and teaching functions of technology transfer. In addition, agricultural extension provides the communication link between users of agricultural technology and the people involved in initiating or directing its development (such as policymakers, planners, and researchers). *Technology supply and service* is concerned with the two other functions of technology transfer. One is the production of agricultural technological components and their distribution to potential users; the other is the provision of complementary services required for their efficient use. In the ten countries studied, the responsibility for the functions of agricultural extension and for all or part of the other functions of technology supply and service was often given to the same organization. Within such an organization, the emphasis on each function tended to vary depending on the clients' demands, the manpower capability within the organization, the type of service expected by the government, and the state of technology development in the country.

Organization of the Report

This introductory chapter is followed by an analysis of the Bank's experience in supporting R&E. The analysis is presented in relation to different levels in the decisionmaking process within national R&E systems. In Chapter 2, the Bank's support for R&E is reviewed in the context of national development objectives and policies. Chapters 3 and 4 deal with the separate reviews of the Bank's support for technology development and technology transfer. The focus, in each case, is on the contribution the Bank's support has made to improving such aspects of national technology development and transfer as organization, management of resources, linkages between research and extension, and the integration of these linkages with other agricultural development activities. The structure of the analysis in the report is to first state the issue as it was identified in the ten countries studied, then to indicate what the Bank did to help countries deal with that issue, and finally to assess the strengths and weaknesses of the Bank's actions.

2

Research and Extension in a Development Context

THE BANK HAS PROVIDED SUPPORT for agricultural R&E to countries with varying socioeconomic and political systems and with different levels of economic development. The development of R&E systems in these countries was dictated by historical factors, some of which continued to be relevant at the time the study was undertaken. In India, Indonesia, Kenya, Mali, Morocco, Nigeria, and Sudan the research function was started prior to independence to serve economic interests primarily outside the country; it was only a coincidence that development interests of the country were served. In all countries in the sample, the R&E functions were started on an ad hoc basis, with the result that responsibility for the two functions tended to be fragmented among several agencies; these functions were often unrelated and uncoordinated. In the ten countries sampled, administrative changes have taken place on a piecemeal basis and, therefore, have not allowed the evolution of institutions responsible for complementary functions alongside R&E institutions.

In many countries, the capability to plan and monitor the use of development assistance is limited. This limitation, coupled with the multiplicity of agencies involved in development assistance, has often resulted in a number of approaches being tried simultaneously, or in sequence, to strengthen R&E. Frequently, these approaches have ignored the historical developments and socioeconomic characteristics of the country concerned.

It is against this background that the ten countries sampled had attempted to set up or strengthen their national R&E systems. In addition, they had to overcome several handicaps, the most prominent ones being: (1) the absence of a national objective for R&E and a clear commitment to achieve that objective by policymakers and planners; (2) the inadequacy of the national institutions selected for planning and giving a direction to the R&E

function; and (3) the absence of a mechanism and a capability to provide policymakers with evidence to back up their support (or lack of support) for R&E and to relate the input into R&E with other inputs in agriculture.

National Objectives and R&E

Experience in the ten countries studied indicates that the more successful R&E programs were those that were compatible with national priorities and were based on the country's comparative advantage in producing a particular commodity supported by R&E. To serve as a guide for developing R&E programs, priorities need to be well defined. For example, a commonly stated priority in the development plans reviewed was self-sufficiency in food production. For R&E, this priority can have operational meaning only if it is further defined to make clear the means available to the country for achieving this goal, the specific foods needed in the country, and the varieties of foods the country could and should produce.

Often, however, development plans lack detail in the objectives and priorities specified. R&E technical personnel in the countries visited complained about the lack of resources to do an efficient job. The lack of policy guidance and support, in their opinion, frequently created a situation in which R&E personnel were expected to cover all aspects of agricultural production and all crop and livestock activities. Consequently, the resources available had to be shared among all the research activities they perceived to be needed.

These observations are supported by the imbalances observed between the institutional strength of national R&E systems, the financial resources of the countries, and the adequacy of the physical infrastructure, on the one hand, and the number of R&E activities being carried out, on the other. Two points are often mentioned:

- The element of uncertainty in planning. The uncertainty is caused, in part, by the lack of appreciation by policymakers and planners of the resources available to the country and of the chances of their continued availability over time (particularly of external funds). This is supported by the observation that most of the plans reviewed provided alternatives for use of resources (in R&E as well as in other development activities) that may eventually become available, rather than a plan of action for achieving specific objectives for which resources were either identified or a strategy prepared to mobilize them.

- Insufficient confidence in the R&E system. While planners and policymakers recognize the beneficial role of R&E in agricultural development, the lack of concrete results from R&E programs to date has apparently led them to withhold strong support for R&E services.

There were, of course, exceptions to the lack of clarity in objectives and priorities mentioned earlier. These exceptions involved cases in which governments made determined efforts over a period of time to increase the production of certain crops. The most outstanding examples were those involving wheat in Turkey and India, rice in Indonesia, corn in Kenya, rubber in Thailand, and soybean and sugar in Brazil. These exceptions involved crops that policymakers and planners recognized as priority crops, and they effectively committed their countries (including their R&E organizations) to supporting increased production of these crops. Experience in these cases indicated that having a clear national purpose to which the country is committed was a principal element in mobilizing and motivating even weak R&E services to be responsive to the country's needs.

The Bank's perception of a country's development objectives and its priorities is normally expressed in the Bank's economic and sector reports. Concerning the ten countries included in this study, these reports have indicated a broad agreement between the Bank and the governments on development objectives in the agriculture sector. There have been, however, differences in the emphasis given to various objectives. In Brazil, for example, the objective of increasing production has appeared in documents of both the country and the Bank. The government documents have emphasized the production of export and fuel crops, while Bank documents have emphasized increases in the production of crops that would enhance equity in income distribution and that are grown in areas that had received less attention in the past. These differences do not necessarily reflect opposing points of view; they reflect instead the fact that the Bank has placed emphasis on those areas more directly related to its lending program as agreed with the country.

As in the case of national development plans (although to a lesser extent), development objectives formulated in the Bank's reports on the agriculture sector and on the overall economy were too broad, which made it difficult to assess the relationship between the Bank's support for R&E and its support for a country's development goals in the agriculture sector. In addition, the objectives often related only marginally to the means available to the country (including funds from the Bank) to achieve them; this made it even more difficult to assess the long-term significance to the country of the Bank's support to its R&E. The Bank's documents, however, have usually identified priority areas for lending by the Bank. These have included, for example, increased production on small farms, intensification and horizontal expansion of cultivation, rain-fed or irrigated agriculture, and production for local use or for export.

The identification of these priority areas has enabled the Bank to focus its own support for R&E on selected crops, population groups, or regions. But this step has not been emulated by the countries included in this study for at least two reasons.

First, priorities identified in the Bank's reports were not always consistent with the social and political realities in those countries. Under such circumstances, some governments gave tacit priority to those development activities for which outside support was likely and thus provided a favorable environment for foreign development assistance. Simultaneously, they devoted whatever other resources were available to development activities that, although not of top priority from a long-term development perspective, nevertheless allowed the government to face up to the country's immediate social and political constraints.

Second, most of the Bank's loans were made for specific projects. The Bank found it difficult to tackle economywide issues in a project context and, therefore, preferred to rely on broader policy dialogues with the countries. But because of the shortage of qualified and experienced manpower in several countries, the Bank had to rely on its own economic and sector work and its limited staff to cover various development issues and to help member countries find practical solutions. In R&E, the contribution of the Bank's economic and sector work in designing a strategy for strengthening these two functions has been limited.

Organization of National R&E Systems

Historically, the allocation of responsibilities for R&E among organizations in the countries studied followed the same institutional structure used for the administration of production and trade activities. For example, if a commodity was produced primarily for export, the responsibility for research and extension for that commodity would often be assumed by the agency overseeing its production or its export. (Rubber in Thailand, citrus and vegetables in Morocco, and cotton in Mali are examples.) If an agricultural commodity was the basis for the growth of an agroindustry in the country, responsibility for research and extension for it would be assumed by the agency overseeing such activities. (Sugar, coffee, and tea in Kenya are examples.) Similarly, if a commodity was of primary importance to rural people, the responsibility for R&E would be assumed by an agency overseeing rural development (such as the offices of agricultural development in Morocco and of agricultural development projects in Nigeria). The functions of research and extension were divided further to serve the needs of food crops, livestock, forestry, horticulture, rainfed and irrigated agriculture, commercial and subsistence agriculture, and privately supported, publicly supported, and parastatal agriculture.

Some of these subdivisions of R&E functions or their variations have been tried in every country included in this study. Attempts have also been made to unite these functions—at least administratively—in every country, which testifies to the widespread awareness of the disadvantages of fragmentation. In contrast, some of the reunited systems have been allowed to separate, out of recognition that while grouping the functions under one administration may be useful, it can be only part of the solution to providing adequate R&E services.

As early as 1929, policymakers in India established the Indian Council of Agricultural Research (ICAR) as the apex organization for education and research on crops and animal husbandry in the country. ICAR was made responsible for planning research and for coordinating its implementation nationwide. The council enjoyed a semiautonomous status and was given control over most of the funds allocated for agricultural research. As a direct link between research and policymakers, the director general of ICAR was also made the secretary to the government of India for the Department of Agricultural Research and Education.

Responsibility for the extension function, which until the mid-1970s was shared between the state Departments of Agriculture and the local community development administration, was shifted wholly to the state Departments of Agriculture. Coordination among the states became the responsibility of the Directorate of Agricultural Extension, an agency within the federal Ministry of Agriculture. A major purpose of these changes was to give prominence to the extension aspects of technology transfer. In contrast to ICAR, legal constraints prevented the Directorate of Agricultural Extension from getting control over national funds allocated to extension. Moreover, the directorate's role in planning and in coordinating extension activities remained limited because the agency lacked the authority and the means to evaluate and control such activities.

In Brazil, as a result of policymakers' recognition of the need for intensifying agriculture, the Brazilian Agricultural Research Corporation was established in the early 1970s as a semiautonomous agency to plan and coordinate national research. A decision was also made to undertake a massive training effort, both in Brazil and abroad, and to establish strong links with outside research institutions. Because of the absence of an effective research program to generate relevant technological practices, early efforts to build a national extension system concentrated on the supply and services aspect of technology transfer. This was later corrected, following the reorganization of the research service.

The extension system was reorganized into a national semiautonomous corporation under the Ministry of Agriculture; it was made responsible for the formulation, coordination, and monitoring and evaluation of existing programs and policies. At the state level, extension organizations were made responsible for the implementation of extension programs. In contrast to India, the reorganization of the extension service has not resulted in a formal institutional separation of the functions of supply and services from the functions of extension.

In Indonesia, Sudan, and Thailand, the early stages of development of R&E systems reflected the historical orientation of the organized section of their production systems to supply raw agricultural materials (such as rubber, cotton, and other commercial crops) to industries outside these countries. Policymakers in these countries had to go (or are still going) through a process of trial and error to expand and adjust their old systems to serve the needs of the entire agriculture sector.[1] Repeated changes in the organization of R&E in Kenya, Morocco, Nigeria, and Turkey were indicative of the awareness of policymakers that their systems had not been adequately responsive to national needs. One observation, common to these countries and others, was that attempts to reorganize the R&E system were not comprehensive because they did not address the major factors (such as manpower, operating funds, allocation of responsibilities, authority, and staff motivation) that accounted for the weakness of the existing organization. In addition, policy decisions related to organizational reforms often addressed the consequence of the problems (such as ineffective organization) rather than the problems themselves (for example, limited manpower capability and limited policy priority given to R&E).

The persistence of weaknesses in the national R&E systems of these countries allowed a demand for technology to grow unsatisfied. As a result, some R&E clients sought services outside the national system (these included agencies responsible for the implementation of rural development programs) or produced their own services, often independently of the national system (the research on fruits and vegetables in Morocco, for example). Because of such reactions by the users of R&E services, there has been an increase in the awareness of policymakers of the usefulness of R&E, of the inadequacy of national R&E organizations, and of the

1. Appendix 1 provides further details.

need to rethink their approach to building a permanent national R&E capability. There is evidence of that rethinking in the nature of actions taken by policymakers: a progressive change in favor of planned and integrated interventions in some countries (Brazil, India, Indonesia, Sudan, and Thailand, for example).

The diversity observed in the national R&E systems and the frequent changes made in their organizations raise the question whether the Bank had any role in bringing about these changes to help countries adapt the organization of their systems to their needs. On this issue, the countries included in the study can be divided into two groups: one comprising Brazil, India, and Indonesia, and the other, Kenya, Mali, Morocco, Nigeria, Sudan, Thailand, and Turkey.

In Brazil and India, and partly in Indonesia, the Bank has contributed directly to helping the countries reorganize and strengthen their R&E systems. This was done by supporting full-scale research or extension projects. In the other seven countries, the Bank was faced with weak national systems and limited response by policymakers to suggestions for formulating and implementing the policy actions necessary to strengthen R&E. In the latter countries, with the exception of Thailand, the Bank has limited its support for R&E to components in development projects, many of which have been formulated and implemented independently of the national R&E systems.

The primary tool traditionally used by the Bank to discuss policy issues related to national R&E systems has been the project appraisal process (and to a lesser extent, the project preparation process). The adequacy of these processes as a conduit for policy dialogue has depended on the extent to which the country has effectively participated in them. Where the Bank worked on the proposals initiated or supported by the country (as in Brazil, India, and Indonesia), its input in the policymaking process was generally positive. Where the country's input in the preparation and appraisal processes was weak (either because the input was limited or because the input did not come from people who were able or authorized to tackle policy issues), the impact of the Bank's support was slow to materialize. Kenya, Mali, Morocco, Nigeria, Sudan, Thailand, and Turkey are examples of the latter case. Even though the impact of any single project was, as expected, limited, the cumulative impact of successive policy interventions was considerable in some countries. In these countries, the Bank's limited involvement in supporting research or extension components has already led either to the extension of its support to the whole national system (as in Thailand and Sudan) or to an ongoing active dialogue on how to proceed in that direction (as in Morocco and Turkey).

Project preparation and appraisal processes were used as the primary tools for policy dialogue on R&E because economic and sector reports failed to provide a basis for it. The coverage of R&E in the reports reviewed was generally sketchy and, in most cases, provided little guidance to staff or consultants engaged in deciding whether and how the Bank should provide support for R&E in a given country. The reports provided even less guidance as to whether the Bank should support or discourage policy decisions affecting R&E in a country. Even when specific recommendations were made, they often came out as isolated judgments rather than as results of a detailed analysis of the R&E services in the agriculture sector.

In conclusion, the Bank's experience as reflected by the ten countries included in this study has indicated a need for more effective input from the country in the preparation of R&E components or projects. There is also a need for more analysis to relate project proposals to the development environment in the country and to relate Bank-supported activities to other agricultural development activities.

Several of the country officials interviewed on the preparation and appraisal of R&E components or projects felt that the Bank tended to impose a certain form of organization for research or extension on the country. These officials argued that consultants who prepared R&E components and projects on behalf of governments tended to concentrate more on finding out what ideas were "in vogue" in the Bank than on studying the local situation and discussing with local people how best to address the constraints they faced. In these officials' view, consultants often started their work with certain R&E organizational structures and methodologies already fixed in their minds. The preparation work thus tended to center on identifying ways to adapt the local environment to fixed parameters, instead of on selecting an organization and a methodology that would best fit the local environment. This point of view was presented with particular emphasis in Indonesia, Morocco, Sudan, Thailand, and Turkey. In Indonesia, the criticism was applied to the extension organization only.

Whether such a statement was right or wrong cannot be unequivocally demonstrated. From an operational point of view, however, the fact that such a perception of the Bank's role existed in member countries deserves mention and careful consideration by the Bank's staff.

Resource Allocation

The issue of resource allocation to R&E was often viewed in mainly quantitative terms. Knowledge of the performance of past R&E programs and of the actual needs of R&E clients was limited. Expectations in terms of results to justify the allocation of resources to R&E were not clearly stated. The policymakers and planners interviewed were unanimous in the opinion that not enough resources were allocated to R&E; this view was broadly held in all the countries studied, despite the variations in the levels of R&E expenditures, in the quality of use of the resources already allocated to R&E, and in the concern of policymakers for obtaining additional resources as opposed to making productive use of existing R&E resources.[2]

Resources were also often allocated with little regard for the long-term needs of R&E. Policymakers often reacted to the availability of external funds rather than seeking them as part of a planned effort to supplement local resources. Moreover, a high proportion of these external funds was normally used within a

2. As a proportion of the value of agricultural production, research expenditures alone ranged in 1980 from 0.2 percent in Indonesia to 1.2 percent in Brazil. The rates for the other countries were 0.26 percent for Thailand, 0.29 percent for India, 0.3 percent for Morocco, 0.4 percent for Turkey, 0.6 percent for Sudan, 0.7 percent for Nigeria, 0.8 percent for Mali, and 1 percent for Kenya. These rates put half the countries included in the study ahead of the U.N. World Food Conference's suggested target of 0.5 percent of agricultural domestic production allocated for research by 1985.

given period for the development of infrastructural facilities, the services of consultants, and, occasionally, the training of staff. At the end of the period, countries often were unwilling or unable to make adequate provisions in the budget to continue operating facilities built with short-term outside assistance. Sometimes the countries did not have qualified manpower to replace consultants and run the facilities productively. The result was often an added liability to the country because of sunk capital and civil servants with no alternatives for employment.

The assistance the Bank has offered member countries to improve their processes of resource allocation has varied from country to country. One means used by the Bank has been broad-based support to national R&E systems. Because the Bank's participation is always dependent on local participation in the project, the Bank has sought to influence resource allocation to R&E from within the country. In the countries studied, it was not always clear that any increase in such an allocation was sustainable. For example, in Sudan and Thailand, where the issue of resource allocation has been critical, the Bank's support created additional demands for resources that were beyond the countries' capacity to meet, given their existing priorities. In Brazil, India, and Indonesia, however, the Bank's support has contributed both to an increase in local allocation for R&E and to an improvement in the use of resources already allocated.

Another means used by the Bank has been cofinancing; this has been used in *full-scale* research or extension projects in Sudan and Thailand. In these countries, cofinancing has been successful in bringing about an increase in the funds allocated for research; the amount has been substantially larger than what the Bank could have reserved for them. More important, however, was the fact that cofinancing enabled policymakers to adopt a unified approach in the use of resources to strengthen their national research systems. Cofinancing with outside agencies of projects with research or extension *components* was more frequent, although the amounts involved were small in relation to total project costs. In addition, the components addressed specific needs of projects often in isolation of national R&E systems.

In the projects reviewed, the Bank's support for R&E started in the form of components in agricultural and rural development projects. Subsequently, Brazil, India, Indonesia, Sudan, and Thailand obtained the Bank's support for full-scale research or extension projects, or both. Turkey and Morocco have also been discussing with the Bank possibilities for full-scale projects. Thus, the Bank's support for R&E in the form of components in agricultural and rural development projects was instrumental in keeping the dialogue going between the Bank and its developing member countries on how to proceed to strengthen R&E.

Many of the officials in developing countries who were interviewed, however, pointed out two shortcomings in the dialogue. One is that the developing countries did not take a strong lead in guiding the Bank toward areas where its support could best help them. The other is that policymakers were often more impressed with the novelties of organization, methodology, and hardware of the R&E system than with its effectiveness, with the result that, in many instances, countries tended to allocate resources to introduce new systems rather than to make better use of, and progressively build on, existing systems.

In summary, the Bank has been more successful in helping countries address the quantitative aspect of resource allocation than the qualitative aspect. This emphasis may be a result of the fact that the quality of resource allocation is affected by sectoral and macroeconomic policies not normally discussed in the context of R&E projects. (These include policies on employment, incentives and human relations, allocations of resources to complementary services, and coordination of outside assistance.)

In balancing support for R&E against that for complementary services, the Bank has not done enough to encourage—directly through its own lending program or indirectly through continued discussions—the coordination of support for services, the effectiveness of which is enhanced when the timing and the extent of their availability are synchronized with R&E development. Obviously, the Bank could not be expected to provide support for all services that complement R&E. The design of Bank-supported projects, however, could benefit from a better assessment of other services, on which the effectiveness of R&E services would depend.

With regard to the continuity of flow and productivity of resources allocated to R&E, the Bank has consistently supported the principle that national budgets should permit support to a project after the investment period. While this principle was reflected in the Bank's appraisal documents and in legal contracts with borrowers, it was limited to individual projects and did not reflect a proper assessment of the aggregate demand on available resources by various development activities in the country (including those activities supported by other external donors). The case studies showed a need for a closer relationship between member countries' available resources (both present resources as well as those that could be generated in the process of development) and the extent and timing of their commitments to investments such as those in R&E that require public resources over a longer period.

Allocation of Resources between R&E

The countries studied, like other developing countries, spent considerably more money on extension than on research in the 1960s. However, in higher-income countries, the investment in research increased faster than the investment in extension during the late 1960s and early 1970s. In countries with low incomes ($150 or less on a per capita basis), the expenditures on extension continued to rise faster up to 1974. The tendency of poorer countries to invest more heavily in extension than in research was a reflection of the view of some policymakers that quicker gains in productivity could be achieved by direct transfer of imported technology to farmers through extension services than by the lengthy process of developing indigenous technology through local research. This perception was heightened by the widespread shortage of trained agricultural scientists, particularly those qualified to conceptualize, plan, and manage research effectively.

A negative consequence of inadequate allocation of investment funds between extension and research and their improper sequencing is the scarcity of well-adapted technology for the extension services to offer farmers. This scarcity slows the growth of agriculture, especially of food production in many countries, and lowers the effectiveness of extension workers. A positive by-product of this situation, however, was the establishment in 1971 of the Consultative Group on International Agricultural Research (CGIAR) and, through its support, the strengthening or creation of thirteen international research centers or programs. These centers were set up largely because it was felt that there was a need to reinforce weak national research systems. The intention was to assist national centers to improve the research methodology, to facilitate the sharing of information, and to train staff to make national systems more productive.

While the recent actions taken in support of research suggest a growing awareness of its importance, they still offer little guidance on what a desirable balance in investment between research and extension ought to be. Similarly, there is no empirical evidence that can throw light on what the optimum allocation of resources ought to be between the two functions. There is, however, ample evidence to show that investments in support of either function have been useful only to the extent that the other function could perform simultaneously at an acceptable level of efficiency.

The Bank has not been consistent in balancing its support for R&E. In some cases, its support has improved the balance overall; in others, it has perpetuated or aggravated existing imbalances. Brazil, India, and Indonesia have relatively strong research services; Brazil and India have also acquired fairly strong extension organizations, although for various reasons their quality has lagged somewhat behind that of their research counterparts. Indonesia had a very fragmented extension system at the time of review. The Bank's support was fairly well balanced between R&E in all three countries. In Mali, Morocco, Nigeria, Thailand, and Turkey—countries in which research tended to be weak because of lack of resources or because of poor organization and management—the Bank gave stronger financial support to technology transfer than to research. In Sudan, where extension was extremely weak, the Bank provided stronger support to research than to extension.

A proper balance in investments is required between research and extension. This balance has to be decided by the country through a careful and systematic review of the machinery for technology development and transfer in its entirety. While the need for a review was recognized, such an undertaking was often inhibited by the fact that research and extension functions were physically, administratively, and psychologically separated. The identification, preparation, and appraisal of projects often suffered from the absence of such an overview in the ten countries studied.

Information in project documents indicated that extension projects or components were often prepared and appraised with limited or no reference to the research capability available and without specific reference to the technologies to be transferred.

Similarly, research projects or research components were often appraised with no reference to the mechanisms available for making research responsive to perceived needs and for transferring the research results to clients. Supervision reports of extension projects or components have often commented on the absence of technology to extend, but have stopped short of recommending specific corrective actions because they would have fallen outside the scope or time span of the project.

3

Technology Development

THE FUNCTIONS OF TECHNOLOGY DEVELOPMENT in the countries included in this study have generally been carried out by several agencies. The primary responsibility has usually been given to an agricultural research organization, while lesser roles have been assigned to training and other technical agencies. The functions of technology development have been performed independently of those of technology transfer. For this reason, this chapter addresses only issues specifically relevant to technology development. Five topics are discussed: the allocation of resources within research; the development and management of research resources; the organization of research; the approach to research; and the linkages between organizations involved in national technology development and those dealing with international research, and between technology development and technology transfer among national organizations. These topics focus on the situation in the countries included in this study, the Bank's perception of issues related to such topics, and the Bank's ability to assist its developing member countries in dealing with these issues.

Resource Allocation, Development, and Management

Among the ten countries studied, Kenya, Nigeria, Sudan, and Turkey appeared to have reasonably adequate resources to make an impact on major production problems, if the resources were properly focused and managed. In contrast, research services had a limited impact in these countries; this must be rated as one of the most serious weaknesses in research management. The issue of resource allocation has to be viewed from three angles: resource allocation, resource development, and resource management.

RESOURCE ALLOCATION. Resources were frequently allocated to crops or agricultural problems of less than top priority; this left key agricultural commodities with inadequate research. In Sudan, fruits and vegetables were not considered key crops for the local population or for export, nor were they targeted in the national plan for further development. Yet, a third of all research activities in Sudan involved fruits and vegetables. In Turkey, even though livestock provided about 35 percent of the gross value of agricultural production, animal research received only 8 percent of scientific resources and 13 percent of expenditures on agricultural research.

Resource allocation within research was not very responsive to changing social, economic, or agricultural conditions. In Sudan, cotton dominated the research scene, particularly on irrigated land. Very rightly, cotton received a major share of research resources. However, the country's desire to diversify and intensify irrigated crops by increasing the production of wheat, sugarcane, and groundnuts was not translated into the sufficient scientific manpower necessary to help achieve that goal. Another example, which applied to all the countries covered, was the tendency to respond to new challenges by adding new programs or new institutions, rather than by reallocating resources within the existing system and phasing out redundant programs. Certainly, closing down programs may be politically difficult, but at least an effort could be made to limit the creation of new programs and to reorient existing ones.

The regional emphasis in research tended to favor the advantaged regions. This can often be justified, particularly at the early stages of development, if it is not done to the complete exclusion of the less advantaged areas. In Indonesia, Kenya, Morocco, Sudan, and Thailand research was focused on developing areas with a high agricultural potential or with irrigation. However, these countries and others were running short of undeveloped areas with high agricultural potential or of land near existing infrastructure. Past resource allocation patterns in Brazil, Indonesia, and Nigeria did not provide for research to extend the frontiers of cultivation to new and often less favorable areas with difficult soils, new crops, and new agricultural practices; nor did they prepare for the handling of social issues, such as those related to population migration.

The gaps in priority research areas were often attributed to fragmentation of research among ministries and departments. These gaps were widespread, particularly where several disciplines had to interact. The most important and most common gap concerned ruminant livestock production, which involved crops or pastures, animal husbandry, and veterinary medicine. Not only did several countries have separate ministries of agriculture and animal production, each with their own research organizations, but work on pastures and fodder crops was often undertaken at stations that concentrated solely on plant production. Forestry also had its own research organization in several countries, and this further complicated the task of land resource management, particularly in fragile environments in the drier and more mountainous areas of India, Indonesia, Kenya, Mali, Morocco, Thailand, and Turkey.

Another reason for the gaps was the inability of many research systems to mobilize the available manpower to serve the country's needs. Often, the researchers' training backgrounds were the only determinants of the nature of their research work; this was due to the frequent absence of local capability to help researchers adapt the use of their skills to address new research problems or to address the same problems in different environments. The researchers, thus, often became the raison d'être of the research. Moreover, in many countries, the geographical distribution of research facilities and the quality of work conditions were not conducive to attracting qualified researchers to work in areas where research was most needed.

RESOURCE DEVELOPMENT. The study showed a lack of synchronization between institutional and physical resource development. The gap existed between the development of manpower and the construction of research facilities. In the countries visited, research stations were underutilized or unutilized because of the lack of qualified manpower or of the means to make use of the existing manpower and physical structures (such as living accommodations, research equipment, and mobility). The lack of qualified manpower was particularly evident in Indonesia, Mali, Morocco, Nigeria, Kenya, Thailand, and Turkey. In these countries, the scarcity of research skills was exacerbated by the dispersion of researchers to a large number of research stations. Only in Brazil and Indonesia was a deliberate decision made to go ahead with the development of physical research facilities so that manpower, which was simultaneously being trained, could be readily absorbed and made productive at the end of the training cycle.

In the other countries, the lack of synchronization was mainly the result of history. Very little emphasis had been placed on the development of local manpower. The shortage of capable and experienced people could be ended only with good planning and the passage of time. In the interim, developing countries would need to place more emphasis on manpower development. This step would require systematic assessment of the supply of qualified people as part of the appraisal of each research project.

In matters of developing institutional resources, the emphasis on manpower development was weak while that on improving the organizational structure was strong. But even the stronger organizational structure, so far, has not led to any major improvement in research management. Indonesia was the only country that had in its existing research entity specific objectives concerning manpower development, organization, and research management and targets concerning the timing of their implementation. In Brazil and India, substantial progress had been made prior to the Bank's support in adapting the organization of research to the countries' needs and in developing manpower. Research management, however, remained in its early stages of development. The other seven countries were at an early stage in dealing with all three aspects of institutional resource development.

When there was a choice between the rehabilitation of old physical research facilities and the development of new ones, the countries studied often favored the latter option. Many officials in developing countries associated this preference with the nature of "tied funds" coming from outside the country. Such funds were often channeled into building new research facilities, while existing facilities were underutilized (often because of limited operating funds and staff); no decision was made to rehabilitate or abandon the existing facilities.

RESOURCE MANAGEMENT. Research administrators frequently mention that they lack financial resources (operating funds) and have difficulty attracting and retaining qualified staff. Coincidentally, money and staff appeared to be the least effectively managed resources in the research organizations visited. Funds were allocated to research with little bearing on the efficiency of research management, although lack of funds might have seriously reduced the scope of research in many cases. Many of the research administrators interviewed indicated that their discretionary authority was limited in two areas: One was in the allocation of funds within the research system's budget (in both capital and operating expenditures). The other was in the hiring and firing of staff and in providing the means to motivate good performers.

Research administrators in India, Kenya, Morocco, Nigeria, Sudan, Thailand, and Turkey stressed the difficulty they had attracting and retaining good research scientists. In the view of the study team, this difficulty was the result of limited financial and professional incentives, the lack of a career structure, poor work conditions (including a lack of funds to undertake meaningful research), and legal limitations on the management responsibilities of administrators. Sudan provided a striking example of how poorly resources were managed in a research organization. Even though the country has well-trained research scientists—among the best available among the countries studied—as a group, the research scientists in Sudan were among the least motivated.

Project documents make it clear that the Bank was aware of many of these problems. How the Bank addressed itself to the problems of resource allocation, development, and management and what the extent of its impact was are illustrated in the following paragraphs.

First, the Bank made funds available to augment resources available for research in general, and for the development of physical research infrastructure in particular. In Brazil, India, Indonesia, Sudan, and Thailand, where the Bank funded full-scale research projects, its support was concentrated on the development of physical and institutional resources. In Kenya, Mali, Morocco, Nigeria, and Turkey, where the Bank funded research components as part of agricultural and rural development projects, the emphasis was placed on the support for undertaking local-level research work and on physical infrastructure development.

The Bank's support for research was perceived by research administrators as a factor contributing to a more balanced allocation of resources within national research systems. The balance applied both to the geographical coverage and to the crop coverage of research.[1] In general, the Bank was perceived as having had a positive impact on resource allocation within research systems, even when its financial input was small. This perception led to an increase in the demand for Bank resources in support of research. The Bank, so far, has responded positively to the increase in the demand for funds in Brazil, India, Indonesia, Sudan, and Thailand. In doing so, the Bank overestimated in some cases the member country's ability to sustain such expanded support for research through internally generated resources.

In appraising future research projects, it would be meaningful if a country's capability to sustain research through domestic resources was assessed in the context of agricultural sector work rather than as part of a project appraisal. In Kenya, Mali, Morocco, Nigeria, and Turkey—countries in which the Bank's support for research remained limited to components—the officials who were interviewed supported the idea of linking such components to the country's national research service. This linkage would require a more intensive appraisal and supervision of those components than has been done by the Bank in the past.

Second, the Bank consistently supported the development of physical research facilities. This support reflected the country's requirements whenever Bank-supported projects dealt with the national research organization (as in Brazil, India, Indonesia, and Thailand). In those cases in which the Bank supported research outside the national research organization, the development of physical research facilities often did not take into account existing facilities (often because they were poorly operated). In many instances, this lapse resulted in the duplication of facilities (as in Kenya, Mali, Morocco, Nigeria, Sudan, and Turkey). This apparent contradiction in the Bank's approach to physical research facilities reflected, to a large extent, the countries' readiness to address weaknesses of their national research systems.

In countries that showed such readiness, the Bank's support was reasonably well integrated into each country's overall support for

1. Appendix 2 provides comments on the Bank's role in supporting resource development and management in specific countries.

research. In countries that did not show such readiness, the Bank sought to help keep the development process going through self-contained investment programs, including, when absolutely necessary, a limited input for research. In the latter case, the duplication of existing research facilities, although undesirable under normal circumstances, could not possibly have been avoided, unless the Bank decided either to provide no development support or to help the countries concerned find and implement programs to systematically address weaknesses in their national research organizations. Substantial additional resources would have been required to simultaneously eliminate weaknesses of other services.

Concerning the management of physical resources, documents of both the Bank and member countries were noticeably silent on the management of research stations (managing the stations' resources and synchronizing their availability for undertaking planned research) or assumed that such a function was synonymous with research management (planning, programming, directing, and monitoring research).

In the development of institutional resources, the Bank's support generally focused on organizational changes and, to a lesser extent, on manpower development and improvement in researchers' working conditions. Support for upgrading the organization of research was limited to those countries in which the Bank financed full-scale research projects (such as Brazil, India, Indonesia, Sudan, and Thailand). National research administrators in Brazil, India, and Indonesia expressed satisfaction with the Bank's support, while those in Sudan and Thailand expressed some reservations. In Thailand, although national research administrators agreed with the principles underlying the reforms supported by the Bank, they found the reforms too drastic within the life of the project because of inconsistencies with national budgeting procedures. These procedures were not properly assessed during project appraisal. In Sudan, administrators of the national research organization felt they had no role in defining or in implementing organizational reforms they were expected to adopt at the end of the project period. Consequently, they foresaw the Bank's contribution to be short-lived.

Manpower development and the improvement of researchers' working conditions were dealt with by the Bank to some extent in the ten countries studied. In terms of numbers, however, the Bank's support for the training of research scientists was limited, except in Brazil and in India.[2] This limited support was particularly noticeable in Kenya, Mali, Morocco, and Turkey—countries with no local capability for higher-level training and with limited access to other sources of funds to pay for the training. In terms of quality, gaps were found in the level of the scientific qualifications of researchers and in the coverage of disciplines needed. In six countries, less than 10 percent of the research staff had training at the doctoral level. In eight countries, more than 40 percent of the research staff had only a bachelor's degree or less.

The 128 projects reviewed in ten countries indicated a reluctance within the Bank to finance formal, high-level training (particularly at the master's and doctoral levels). The financing of on-the-job training or vocational training was favored instead. Such a bias is unfortunate because research is a field that involves considerable conceptual thinking. Among the important gaps noted in the coverage of disciplines were research management, research station management, agro-climatology, plant physiology, plant protection, water management, animal husbandry, pasture and range management, agricultural mechanization, economic and policy analysis, sociology, and communication. The Bank's policy encouraged the integration of technical, socioeconomic, and environmental factors in research. Judging from the research programs reviewed, however, progress in that direction has been slow.

The Bank has supported improvement of the researchers' working conditions. This objective, however, has been difficult to tackle because of the need to avoid imbalances in the quality of working conditions among different sections of the economy. Because policymakers were generally reluctant to disturb the existing equilibrium, the Bank began to support the provision of location-specific or group-specific advantages to improve productivity in the projects it supported and to demonstrate the need for, and potential benefits from, favorable working conditions.

2. In Thailand, while appraising the extension and the research projects, the Bank's project staff were aware of the training needs, but did not provide for them because other donor agencies were planning to provide such support.

This approach was generally helpful in Brazil, India, and Indonesia, where policymakers were responsive. In Morocco, Nigeria, and Sudan, this approach created friction between staff in Bank-supported projects and researchers in the national organization. In Thailand, government officials felt that this approach, if adopted, could lead to imbalances in employment conditions that the country would find difficult to rectify.

In summary, judging from the case studies, the Bank had a positive impact on resource allocation for research and, to a lesser extent, on the development and management of research resources. It helped several countries expand their research services to previously neglected areas and crops and, to varying degrees, encouraged these countries to give priority to the development of research resources and to their proper management. The basic weaknesses in the Bank's approach were the limited analysis of the macroeconomic and social issues in the appraisal of research projects and, in some cases, the limited integration of Bank support in the country's overall program for agricultural research.

Research Organization

Judging from the case studies, efficiency in the organization of research systems was not easy to achieve in smaller countries with simple administrative structures. In fact, it was the larger countries in the sample that seemed to have more efficient organizations. Six factors emerged as common features of the more successful research organizations reviewed:

- A strong central organization with overall responsibility for most aspects of national research and with funds to carry out its responsibility
- Some degree of autonomy from the routine bureaucratic structure of the Ministry of Agriculture
- Good links to the national planners and policymakers despite the organization's semiautonomous or autonomous structure
- Service units—either directly part of, or closely related to, the central research organization—that perform important planning, monitoring, and evaluation functions

- A planned decentralization of regional research
- Active participation in translating research results into recommendations communicable to users.

Although there was no single formula appropriate to all circumstances, it was clear that any coordinating mechanism must ensure that sound links exist among national policymakers, research planners, and clients. If there is an umbrella management organization for research (which seemed valuable for larger countries), it must have control over research funds and be provided with adequate supporting services for planning, monitoring, evaluation, economic and statistical analysis, and financial management. It was the absence of these factors that rendered coordinating institutions impotent and defeated efforts to establish effective national research organizations. It is important that the coordination function not be confused with that of top-down central control. The latter may be as ineffective as fragmentation.

Fragmentation can make it impossible for research to make an adequate impact. The fragmentation can occur both at the national level among ministries and at the field level among stations. The case studies were replete with examples of both. From a planning and management point of view, if the bulk of research institutions were coordinated by one agency, such as EMBRAPA in Brazil or ICAR in India, it would be less difficult to make adjustments among staff or to shift priorities and resources among the institutes than if they were independently controlled by up to ten different ministries or agencies.[3]

The Bank played an important role in helping its developing member countries consolidate, and improve the relevance of, their research organizations. Where the Bank's efforts were successful, they led to more effective regional decentralization of research (not to be confused with fragmentation). The Bank contributed to the building of strong national organizations in Brazil, India, and Indonesia, although for historical reasons the impact of its measures varied with the stage of development of each system. In India, for instance, basic elements of the national system were

3. For further comments on the organization of research services in the countries studied, see Appendix 3.

already in place when the Bank started its support. The Bank's main contribution was to help in building a regional and local research network by strengthening the state agricultural universities. In Brazil, the Bank provided major assistance to strengthening EMBRAPA both centrally and regionally. In Indonesia, it helped support the creation of a national agency and placed major emphasis on training. Basically, its role in all three countries was oriented toward institution building.

In Sudan and Thailand, the Bank attempted to play a similar role in helping reorganize national systems, but, perhaps because of resistance by established agencies and because of local budgetary constraints, progress was slower.

In other countries, although the need to establish, strengthen, or restructure the overall research planning mechanism was equally great, the Bank's input was limited. The main reason was that policymakers in these countries remained reluctant to support activities from which they expected benefits only in the long term, but for which they had to bear costs almost immediately. In these countries, a more project-oriented role for the Bank was the rule, with its support limited to research components in agricultural and rural development projects. In this manner, the Bank has continued its dialogue with countries on matters concerning research and extension. Efforts were made in several countries to persuade policymakers to reorganize national research agencies that were ineffective or to provide enhanced support for training, management, and other essential elements of a sound research organization.

The major difficulty encountered by the Bank in supporting reorganization of research has been to persuade policymakers to allow national research administrators to go beyond the stage of adopting a modified organization chart by giving them the responsibility, the authority, and the financial means to make such a chart operational.

The Bank's input could be made more productive if it is utilized to progressively strengthen desirable features of existing organizations while slowly eliminating less desirable ones, rather than to induce drastic organizational changes. Changes should be defined in relation to the organizational objective being sought and should be planned with due regard to the strengths and weaknesses of other national organizations and the overall quality of national institutions. Moreover, because organizations rarely work the way they are theoretically intended, the implementation of any such planned changes should be closely monitored and their usefulness occasionally reassessed.

The discussion in the preceding paragraphs suggests that both the countries and the Bank should put more effort into appraising the organizational aspects of research agencies. In addition, the Bank should add more technical input into monitoring the implementation of organizational reforms under research projects. These measures will obviously have implications for the resources required for the Bank's supervision of projects and the kind of staff put in charge of supervision.

Approach to Research

It has been the convention to divide research operations into basic, applied, and adaptive (including on-farm) research. Such distinctions are at best semantic and, in some respects, divisive and undesirable. In effect, the operations involved in the development of knowledge and technology represent a continuum, and their usefulness cannot be inferred from their nature but instead from their relevance to the country's needs.

BASIC RESEARCH. The Bank supported basic research mainly through its contribution to some of the work done or supervised by the international centers. Some support to basic research has also been provided in India and Brazil. In general, however, project documents indicated a bias in the Bank against financing basic research. This bias was usually justified on the grounds that Bank-supported research should be oriented toward solving problems. This study contends that there should be no relationship between the nature of research (meaning the level of the activity on the research continuum) and its problem-solving orientation. The problem-solving characteristic is acquired by research (regardless of its nature) as a result of good conceptualization of research problems and adequate planning and programming of research activities.

The question of who should assume responsibility for basic research in agriculture has not been resolved. A policy issue that may have to be faced before long is whether the CGIAR-supported centers, which are mainly working on applied research problems at an advanced level, should devote more attention to basic research. The issue arises because of the limited capabilities of national programs, the progressive exhaustion of conventional technology, and the challenging environmental and energy problems that are arising. The Bank, as one of the donors to the CGIAR, and with its links to the national systems in borrowing member countries and the research institutions in the developed member countries, may have an important role to play in providing guidance on this issue. More urgently needed, perhaps, is for the Bank to ensure, both through the CGIAR Secretariat and through its action in support of national research, that strong links are developed between national institutions and international centers to strengthen the complementarities between them.

APPLIED RESEARCH. Within the framework outlined in the preceding paragraphs, the main part of the Bank's funds was allocated for applied research. Through its support, the Bank encouraged countries to focus on commodity research within a multidisciplinary framework. In principle this was sound, but there were agricultural problems that required a different approach. In India, for example, ambiguities in the appraisal of the national agricultural research project resulted in the state agricultural universities and their associated regional and local stations (which were the main focus of the Bank's support) concentrating on only three groups of food crops: cereals, pulses, and oilseeds. Admittedly these crops were important, but this restrictive approach could defeat the ultimate aim of the project: to produce technology that could help create *farming systems* well adapted to regional and local needs. Although food production in India was a national priority, regional or local needs could, in some situations, require horticultural crops, livestock, or even export crops to receive equal or greater priority in research than basic foods. Similarly, even though two-thirds of Indian agriculture is rainfed, insistence on giving priority to rainfed agriculture could impede the development of optimum land use policies and reduce the potential returns to the massive investments in irrigation made in India. These problems were further complicated by the progressive transfer of responsibility for applied and adaptive research related to water use and management at the state level from the irrigation command area authorities to the state agricultural universities.

There were similar examples of commodity-oriented research in Kenya, Mali, Morocco, Nigeria, Sudan, and Turkey. While such an approach provided a convenient and relatively straightforward base for planning and for resource allocation, as well as for evaluating the impact of research, it was not appropriate for all agricultural situations in these countries. In fact, in many cases this approach did not provide the right technology for the area development projects that the Bank supported.

This study suggests that a flexible approach be adopted toward research, including commodity research when appropriate and farming-system or problem-oriented research when needed. Farming, except in special circumstances, is composed of a mix of activities, each with its own problems. Both the mix and the problems vary with local circumstances. In some cases, a problem (salinity, for example) may be large enough to justify a special research institute. While it is often appropriate for national stations to focus on a limited range of commodities, or even a single one, the commodities usually have to be put together and tested in an appropriate system somewhere between those stations and the clients. The approach being followed in appraising the state-level research system in India could provide a valuable model in guiding future Bank support to national and regional programs.

ADAPTIVE RESEARCH. In countries with advanced agriculture, it is the farmer who adapts as well as adopts improved technology. The public sector research stations and the commercial agricultural sector offer farmers a series of options in which the quality of the technology is controlled through competition among the providers of technology. The farmers modify many of the recommendations to fit their own farming needs. In contrast, in countries with less advanced agriculture where the farmers' capacity to adapt and the private sector's capability to supply technology are still weak, research agencies adapt and supply research results. Even though their role is different, these agencies are modeled along the lines of the developed countries' public institutions.

The Bank has supported what reports refer to as "adaptive

research." If adapted technology is the product of adaptive research, then judging from field discussions, sector studies, and project documents, it appears that the Bank's support was not instrumental in producing much adapted technology. This inadequacy can be attributed to two factors: (1) ambiguities in the assigning of responsibility for adaptive research among research, extension, and area development staff; and (2) methodological uncertainties concerning the conduct of adaptive research. Both factors are important. In India, Indonesia, Thailand, and Turkey, the ambiguity in responsibility resulted from the same research activities' being independently conducted by several organizations within the same ministry and even by several ministries. In Mali, Morocco, and Nigeria, adaptive research was isolated from other national research activities.

Part of the difficulty in adaptive research may lie in the definitions. It may not be clear what constitutes "applied," "adaptive," "trial," "verification," "demonstrative," and "on-farm" research. These terms may appear to have mere semantic distinctions, but in practice they were the source of much confusion. In India, for instance, one Bank paper placed the responsibility for the conduct of adaptive research (which it redefined as "farm trials") with the state-level Department of Agriculture and the day-to-day conduct of such research with the village extension workers, not with the state agricultural universities that the Bank's project supported as the main instrument for local-level research. Apart from the jurisdictional issues, it was clear that the inexperienced village extension workers, untrained in research and analysis, could not, as suggested in the Bank paper, lay out, observe, and record harvests and economically analyze trials in addition to carrying out their usual duties.[4] The results bore out this contention.

Similar problems were observed in Mali, Morocco, Nigeria, and Thailand, where extension staff attempted to conduct adaptive research and demonstrations. In fact, it is *not* essential that all adaptive research be done on farmers' fields. A combination of

4. The Bank's South Asia Regional Office states that the village extension worker is supposed to be supervised by his agricultural extension officer. Moreover, participation by research staff is encouraged.

controlled trials on outlying stations or on government farms with trials on actual farms may give a more reliable coverage of local conditions as long as the work is performed by qualified staff.

Evidence from the case studies indicated that adaptive research was the weakest, most neglected, and most confused aspect of the national agricultural research systems; this was the case whether adaptive research was considered from the point of view of administrative responsibility, organization and management, or operational methodology. While the Bank recognized the importance of adaptive research, its attempts to support it were not well conceived or well organized, as the examples in the preceding paragraphs have indicated. Adaptive research in the countries studied needs to be upgraded and put under the direct control of the main regional station or a research station serving a particular ecological zone in a large country. In a smaller country, it can be placed more directly under central control. The technical inputs in adaptive research should be derived from mainstream research or from the selected import of research results.

Staff doing adaptive research should be composed of career members of the national research system. Whether they should be dispersed across a chain of adaptive research stations, development projects, and government farms, as was often the case, or located more strategically in one place will depend on local circumstances. Judging from international literature on the subject, it appears that clusters of strategically located staff with full mobility and adequate research assistance might be more effective in many situations; this would be the case particularly when dispersed individuals at small stations lack the mobility to visit farmers and to have interdisciplinary interaction in their work. Staff involved in adaptive research should be jointly responsible with local extension staff for the identification of problems and choice of local priorities for research. They should work with extension staff in the selection of sites, layout of trials, maintenance of records, interpretation of results, and translation of results into an extension message. However, the overall responsibility for on-farm trials should lie with the research staff, not with the extension staff, in the same way as the conduct of demonstrations (once beneficial innovations from adaptive research can be identified with confidence) should be part of the technology transfer function.

Monitoring Research

The research function is not completed once the basic, applied, or adaptive research activities are undertaken. Two other major responsibilities of a research organization remain: the monitoring of research activities and the evaluation of their results, as well as the acquisition of feedback from clients.

Concerning monitoring and evaluation, the lack of capable staff remained a basic obstacle to establishing this function in research organizations. Most research services were found to be accountable only to themselves. Often, there were no defined channels for reporting research results and for getting feedback on the usefulness of research. Economic rates of return to research in developed countries were often cited as evidence of its high payoff. Researchers in developing countries, however, need to produce evidence of the productivity of their own efforts. India and Brazil have attempted to introduce monitoring and evaluation in their research organizations. But even in these countries, competence in these fields has been limited. Major training efforts are needed for these activities to become productive.

Monitoring received more attention than evaluation in Bank-supported research projects; this was understandable in the early years of project implementation, when the impact of investments was difficult to evaluate if the essential construction works were not completed, the equipment not provided, the staff not trained, and the systems not properly organized and managed. However, a criticism commonly voiced in the countries visited was that the Bank put too much emphasis on monitoring quantifiable parameters (expenditures, disbursements, and so forth) and too little on monitoring the quality of programs. It was also noted that the monitoring process was too narrowly restricted to the progress of the project under implementation; it neglected parallel developments of interest, for which support was not provided under the project.

There was evidence that the composition of the Bank's supervision missions did not allow for proper monitoring of the quality of research programs. If Bank-financed research projects were to emphasize more strongly the need to use existing resources more effectively, and if it was agreed that research projects could be improved by spreading a given amount of money over a longer time, then the Bank would have to put more resources and different kinds of resources into supervision. The kind of supervision would also probably change over time; the initial emphasis would be on the development of infrastructure, institutional changes, and so forth. By the end of the second or third year, however, new monitoring parameters would be added. The parameters would include the way programs were being planned, the way research was being conducted, the qualifications of candidates for training, and the way the research strategy fitted in with the country's overall development needs. These are the factors that make or break a research system.

Linkages and the Transfer of Research Results

EXTERNAL LINKAGES. A major share of the credit for coming up with the most useful results in agricultural research in the past two decades should go to the international research centers, particularly those concerned with rice, wheat, maize, and legumes. Although many developing countries were represented on the boards of directors of these centers, the primary input in defining the direction and content of their research programs came from researchers in these centers. To what extent should researchers continue to assume primary responsibility for defining the scope of their research programs? This question illustrates the importance of linkages among national research organizations and between international and national research institutions. The Bank's member countries stand to gain a lot from the research of international institutions, and they can save substantially on their resources by helping direct these centers to serve them.

Attempts at establishing operational links between international and national research organizations have so far been initiated mostly by the international centers.[5] Although most countries have some linkages with external research (with international institutions or bilateral institutions), the extent and effectiveness of these

5. Appendix 4 provides country-specific comments on linkages between national research services and international research centers and other national research agencies.

linkages vary. In Brazil and India, the extensive and successful cooperation with outside institutions points up the conclusion that strong national systems are better able to take advantage of such linkages, at least at the working level, than weak ones.

Despite the Bank's role in the management of the CGIAR system, it is not evident from the case studies that Bank-supported development projects have been designed to take advantage of the opportunities offered by the international centers in the training of researchers and in the programming of research. Training could be the medium through which the Bank could further enhance cooperation between research institutions in developing member countries, on the one hand, and scientists in the international centers and in developed countries, on the other.

INTERNAL LINKAGES AND TRANSFER OF RESEARCH RESULTS. The picture in this area is bleak. For instance, the Indian Agricultural Research Institute has not produced a comprehensive report of its scientific activities since 1977. In Sudan, many researchers in the national research organization did not even recognize the need to be accountable for their results or to be responsible for transmitting the results to potential users. Horizontal communications among research institutions were inadequate in most countries because of the fragmentation of responsibilities. Weaknesses in the mechanisms for communication of research results to farmers were evident, even in countries where the Bank supported national extension projects. These weaknesses resulted mainly from inadequate communication between research workers and extension agents. In several countries, both research and extension services transmitted technical information independently and separately to farmers. Examples of duplication, and even of conflicting recommendations, were found in Indonesia, Thailand, and Turkey.

The quality of both the information and the means of its transmission was often poor. In Brazil, communication was specifically recognized as vital to the success of research, and an attempt was made to train staff and to establish units skilled in the reception and transmission of information; this was very much an exception. The importance of communications as an essential element in solving linkage problems was not well recognized by national agencies or, at times, by the Bank. This lack of recognition was evident from the fact that the supervision reports made no mention of the failure of research and extension training centers to utilize adequate and up-to-date equipment, mass media, and teaching methods, even where these were financed by the Bank.

4

Technology Transfer

THE SYSTEM OF TECHNOLOGY TRANSFER, as defined in Chapter 1, can be subdivided into three main components: (1) agricultural extension or the "knowledge transfer"; (2) the supply of inputs such as seeds, fertilizers, pesticides, farm equipment, and animal feed (including their production, procurement, and distribution); and (3) agricultural services, including credit, machinery maintenance, veterinary service, and artificial insemination. Many national extension services tended to concentrate their efforts on such transfer functions as input supply, credit supervision, agricultural data collection, and a variety of other noneducational tasks.

The functions of technology transfer differed from country to country. In some countries, the supply and service functions were performed by the private sector, while that of agricultural extension (knowledge transfer) was performed by the public sector. In other countries, the three functions were provided by a combination of government institutions, parastatal agencies, and cooperatives.

These functions were organized around three basic approaches. Under the first approach, the three functions of technology transfer were largely organized as separate, vertically integrated activities. Separate institutions were given responsibility for each function, while coordination among them was either informal or nonexistent. National systems in Kenya, Mali, Morocco, Nigeria, Sudan, and Turkey were organized along these lines.

Under the second approach, the functions were integrated horizontally within area development programs. Agricultural extension, input supply, and other agricultural services were organized and administered as an integrated system within a specified area or district. For example, all three functions might be carried out through a district farm service center or an agricultural development office, where farmers could expect to get a package of extension support, inputs, and other services. Rural development programs in Kenya, Morocco, Nigeria, and Turkey followed this approach.

The third approach was a vertically integrated production and marketing system. In this approach, an organization working on the development of a specific commodity (such as tea and coffee in Kenya, rubber in Thailand and Indonesia, and cotton in Mali) provided the necessary extension support, input supply, and other services. In addition, the agency generally controlled the marketing; this enabled it to recover the cost of these different services. Commodity-specific systems were generally used for export or for the production or marketing of other high-value crops grown in a more or less contiguous area or region.

The Bank's support for technology transfer did not follow a single model in any country. In Brazil, all three approaches were supported—sometimes in the same area. In Mali, mixed commodity and area development projects were supported. The Bank's approaches can be grouped into four categories. The first included agricultural extension projects. The intention was to develop an agricultural extension organization that focused on the knowledge transfer function. These projects were found in Brazil, India, Indonesia, and Thailand. The second category involved specific supply and service projects that focused, for example, on a national seeds agency or a fertilizer industry, or on developing a capability for procuring and distributing inputs or for providing credit and other services. These projects were found in India, Indonesia, and Kenya. The third category included integrated commodity development projects, in which agricultural extension was prominent as a project component but the major part of the funds was spent to strengthen the supply, service, and marketing functions. Examples included sericulture and dairy in India, cotton in Turkey and Mali, and coffee and sugar in Kenya. The fourth and most dominant category in the countries studied—in terms of the number of projects and years of support—involved area development projects (also called agricultural or rural development projects) in which the extension and the supply or service functions were frequently integrated and the supply or service aspects often emphasized.

This chapter focuses on the key issues faced by developing countries in organizing and strengthening various functions involving technology transfer. Although the main focus is on the transfer of knowledge, the supply and service aspects of technology transfer are, nevertheless, treated in this analysis to identify how they were linked to, or integrated with, knowledge transfer (extension).

The discussion is organized in five parts: status of extension personnel; resource allocation, development, and management; extension methodology; organization of support for extension; and linkages. The chapter first defines the issues as they were identified in the ten countries studied, then indicates what the Bank did to help the countries deal with those issues, and finally assesses the strengths and weaknesses of the Bank's approaches in providing assistance.

Status of Extension Personnel

Agricultural extension has historically been perceived as a function of low status performed by poorly qualified and poorly equipped persons who deal with poor and, frequently, illiterate farmers in remote rural areas. This perception has been supported by facts: low salaries, unclear job description, poor supervision of performance, and poor quality of work by extension personnel.

LACK OF AN ADEQUATE REWARD SYSTEM. Salary scale and allowances were set by the civil service. The provision for improving salaries across the board or on the basis of merit was limited. Prior to the Bank's involvement, there was no indication in the countries visited that any type of reward system had ever been considered—let alone established.

To deal with the resulting problems, the Bank promoted the idea of special allowances as a means of providing financial incentives to extension personnel (and others) in the projects it supported. Economywide issues related to civil service rules and regulations were not addressed, however. While these allowances may have improved the staff's morale and made field-level extension work more attractive during the life of the projects, their long-term effectiveness may be doubtful for financial reasons (budgetary constraints) and institutional ones (discriminatory practices among groups of civil servants). The Bank also attempted to help countries improve the working conditions of extension personnel through its support for improved housing for field-level agents, loans for acquiring transport facilities, allowances for transport, and the acquisition of extension equipment and materials. This support was provided in varying degrees in full-scale extension projects as well as in extension components of agricultural and rural development projects.

JOB DESCRIPTION FOR EXTENSION PERSONNEL. A common feature within national extension systems has been the excessive involvement of field-level extension personnel in the noneducation aspect of technology transfer. For example in India, prior to the Bank's involvement, village extension workers had multipurpose assignments within the Community Development Program; they spent only a limited amount of their time on agricultural programs and even less time on agricultural extension. In Kenya, Nigeria, and Turkey, agricultural extension workers helped run production campaigns that were heavily involved in the provision of inputs and other services. In Brazil, Morocco, and Thailand, administrative responsibilities diluted the role of extension personnel.

The Bank's approach to this problem, particularly in the full-scale extension projects it has supported in Asia, has been to insist that extension personnel work full time on the knowledge transfer component of technology transfer. In India, Indonesia, and Thailand, the Bank strongly supported a separation of extension from other field activities. To ensure that these agents did not drift back to old patterns of work behavior, the duties of extension workers were highly structured by a series of regular visits to groups of contact farmers. These visits involved communicating and demonstrating specific technical practices to farmers. The visits, plus the fortnightly training sessions, so structured the job assignment of extension workers that they could not easily be diverted to nonextension tasks.

Preliminary indications are that the results of the Bank's efforts to implement full-time extension roles have been mixed. In India, full-time extension has been established; in Thailand, agents appear to be caught between the traditional and the Bank-recommended roles. Although the Bank insisted on full-time extension job assignments, in most cases no provision was made for another agency to handle the many administrative, supply, and service duties previously carried out by extension agents. In Indonesia, although the new roles have been implemented, government officials have sought more flexibility in adjusting assignments; they have observed that there is no need for the same intensity of visits during the off-season. (The Bank's staff were perceived to be dogmatic on the issue of regularity and frequency of visits even during the off-season.) In Brazil, progress has been made in developing clearer extension roles for agents, although many continued to assume major responsibility in servicing rural credit and social programs. In Kenya, Morocco, Nigeria, and Turkey, there has been less emphasis on changing the traditional, supply and service assignments of extension personnel. Of course, there was a question of how much new technology was really available for extension in these countries. In Mali, the approach of the Compagnie Malienne pour le Développement des Textiles (CMDT) to extension, which was supported by the Bank, has been successful in balancing the extension and the supply and service roles of extension personnel. In India also, while input supply was no longer to be undertaken by the village-level worker (VLW), other full-time agricultural staff were assigned to this function at the next level, and provision was made for coordination with, and orientation training for, input supply staff.

To summarize, the Bank's approach in helping to introduce well-defined extension roles depended in large part on the type of project being funded as well as on the level of understanding of, and appreciation for, the scope of the agricultural extension function among the Bank's staff and government officials. Where both government officials and Bank staff were convinced that agricultural extension should be a full-time job, clear-cut extension roles were introduced without regard to the type of project. In commodity-based projects, for instance, it was possible to limit field-level agents to extension because it was recognized that the quality of the product was essential for its sale or export; therefore the need for proper use and improvement of technology was accepted. Among the Bank-supported projects that were based on general food crops and livestock, full-scale agricultural extension projects were more successful than extension components in helping countries adopt full-time "extension" assignments for extension personnel. As Thailand demonstrated, however, countries needed to shift the traditional assignments of extension agents to other agencies to avoid having extension personnel caught in undesirable role conflicts.

SUPERVISION OF EXTENSION PERSONNEL. A common problem associated with unstructured, multipurpose job assignments for extension personnel in the field has been poor supervision of their work. Where it was difficult to define the agents' assignments, it was equally difficult to establish criteria to evaluate their performances.

In the full-scale extension projects it supported in Asia, the Bank attempted to introduce regular, systematic supervision of extension personnel. Nevertheless, the Bank's support in helping improve supervision within national extension services has not been limited to full-scale extension projects. For example, in an area development project in Mali, extension work was reported to be regularly supervised, while in a full-scale extension project in Thailand, the supervision of extension personnel was reported to be weak. In Brazil, extension agents were sometimes supervised by staff from other rural development agencies rather than by extension supervisors. In India and Kenya, the quality of supervision of extension work was reported to vary over time.

Although most supervisors were going through the procedure of making supervisory visits, they did not appear to know what their supervisory job entailed or how to perform it effectively. This problem was attributed to a lack of training—which in turn was largely explained by the fact that most of the projects were relatively new.

An important contribution made by the Bank in full-scale extension projects has been the promotion of the monitoring and evaluation (M&E) function as a means of ensuring the effectiveness of extension. Where the M&E units existed, they concentrated on gathering data on the frequency of agent-farmer contacts, on the rate of adoption of recommended practices, and on yields.

QUALITY OF EXTENSION PERSONNEL. The poor quality of field-level staff was a problem common to all countries. There was general agreement among the countries visited that the poorly educated and poorly trained extension staff were not fully equipped to disseminate improved technology to farmers effectively.

The Bank's response to this problem has largely been one of providing short-term in-service training conducted by subject matter specialists (SMSs). Extension workers were trained for one day each fortnight and given the message to disseminate during the following two-week period. In addition, some efforts were made to establish longer-term preservice and in-service extension training. In India, for example, the Bank supported the repair and expansion of a number of old village-level training centers. In Brazil, new agents received fifty to ninety days of training; provisions were also made for some in-service training courses. There did not appear to be any significant effort in any of the countries studied (with the exception of one in-service program in India) to upgrade the educational level of existing or incoming staff.

To illustrate the magnitude of the problem of training and educating SMSs, it is useful to compare the proportion of SMSs with the rest of the extension staff. In North America and Western Europe, the proportion is 19 percent and 18 percent, respectively. Most of these SMSs hold master's or doctoral degrees. In Asia and Africa, the ratio is 6 percent and the best among these SMSs hold bachelor's degrees. Latin America is better off with 13 percent. The training problem, however, has both quantitative and qualitative dimensions.

Efforts to upgrade the technical and professional skills of extension personnel have been headed in the right direction, but have been inadequate. The concept of fortnightly training appears to be a useful short-term strategy for improving the skills and technical knowledge of the poorly educated extension personnel while they are on the job and for disseminating improved agricultural technology among farmers. However, this training approach is no solution to the long-term task of upgrading the skills, the knowledge, and the attitudes of extension staff in the field. In the long run, each country will need to progressively raise the educational level of its field-level extension staff to help its farmers adopt new technology more rapidly. An extension agent with just a secondary school education will not have the foundation necessary to understand farmers' problems and to teach them about increasingly complex forms of agricultural technology. As the quality of extension personnel improves, countries will need fewer staff because quantity can be substituted by quality.

In India, where extension training centers have been built or expanded for longer-term training, the results have been poor. While classrooms and dormitories have been built, the curriculum, instructors, teaching aids, instructional materials, and equipment for practical training have been severely neglected. Even though some resources were provided for these items, the lack of supervision and attention to training needs resulted in resources being reallocated to other uses.

India's experience showed that organizing extension training centers within the Ministry of Agriculture may not be a satisfactory arrangement. When funds become scarce, training is the first item an extension organization reduces or eliminates. Instructors generally perceive extension teaching as out of the mainstream of the extension organization and a dead-end job; they therefore seek transfer out of these positions at the first opportunity. At the same time, every country included in the study needs to have a serious, long-term educational program aimed at upgrading its extension personnel. A long-term solution to this training problem would be to significantly expand the educational capacity and to improve the quality of agricultural schools and universities.

The Bank has paid limited attention to training SMSs in most countries; in Indonesia, only a minimal effort has been made in this direction. In India and Thailand, SMSs were expected to learn on the job by working with research workers and by attending two-day workshops each month. Field visits indicated that some workshops were being conducted, but they were given little priority by the research workers who were to do the training.[1]

Resource Allocation, Development, and Management

The allocation of available resources among various elements of the technology transfer system within countries was affected by

1. A recent change in research leadership in Thailand is reported to have resulted in researchers' giving high priority to SMS training.

several factors. First, the extension (knowledge transfer) element tended, historically, to be combined at the operational level with other technology transfer functions. This confusion was partly a result of the fact that organized extension was an alien concept in most developing countries; not enough time had elapsed for them to recognize the importance of, and need for, an effective extension service. Second, the extension function depended on the availability of knowledge worth transferring. The development of such knowledge was slow in most countries. The extension element thus had no basis for acquiring recognition as a function. Third, the transfer of technology from more advanced countries had created needs for the supply and the distribution of marketed inputs and for accompanying services at a time when the physical and institutional means available in various countries were not sufficient to meet them. Consequently, the extension function, which was less tangible and more difficult to perform, was often sacrificed in favor of other technology transfer functions. Fourth, various technology transfer functions were considered separately for purposes of the allocation of resources, particularly of those resources originating outside the country. These functions thus competed with one another for resources.

Historically, the allocation of resources within the technology transfer system evolved from the almost exclusive emphasis on distribution of inputs to an increased emphasis on provision of services. Only recently has more attention been given to the transfer of knowledge. The observation in a Bank policy paper that "extension" received more emphasis than research in the past may, therefore, have to be moderated because the statement refers to all technology transfer functions rather than to the extension (knowledge transfer) function alone.[2] Recently, resource allocation shifted in favor of the extension function in Brazil, India, Indonesia, and Thailand. In these countries, an attempt has been made at distinguishing the extension function from those of supply and services. In other countries, the level of support provided to extension was often negligible because of the need to provide support for the other functions.

2. See World Bank, *Agricultural Research*, Sector Policy Paper (Washington, D.C., June 1981).

In extension, the central element in resource allocation was manpower resources. The literature on agricultural extension is replete with suppositions about the actual, the ideal, and the maximum numbers of farmers an extension agent can effectively serve. The ratio of extension agent to farmer has been used as one of the indicators of the quality of an extension service and the extent of a country's commitment to extension. Experience in the countries studied showed that merely reaching a particular ratio did not necessarily result in an effective extension service. In fact, the setting of such a ratio as a target often resulted in the misallocation of manpower and in the mismanagement of other resources within extension services. An extension agent to farmer ratio could have meaning only in relation to five parameters: (1) the country's overall demand for skilled manpower and its training capacity; (2) its budgetary means and its capacity to generate funds to support extension on a sustained basis; (3) differences in the level of development of farming among various groups of farmers; (4) the potential for tradeoff between the use of manpower and communication media for extension; and (5) the potential for tradeoff between the use of a few highly qualified and mobile agents and the use of large numbers of poorly trained and less mobile agents.

With the exception of India, all the countries visited gave priority to increasing manpower resources within extension services. The lack of qualified extension agents and the limited national capacities for training led these extension services to sacrifice the quality of manpower to achieve quantitative targets. This concern with the sheer number of extension agents was not balanced with the provision of adequate resources to make the extension personnel more useful, however. Extension agents in the countries visited expressed concern about the inadequacy of resources such as transport, extension materials, and supplies for demonstration. In some countries, extension agents also expressed concern about limited office space, inadequate housing facilities, and the absence of meeting places.

Thus, while some countries have recognized extension as an essential function, they have not been able to make the necessary adjustments in the nature of resources and in their allocation within the technology transfer system. Although extension resources, particularly human resources, may have been suitable for the communication of simple technology, they have remained inconsistent with the requirements of a knowledge transfer function.

The Bank contributed, in varying degrees, to the augmenting of resources available for extension. Its emphasis was on the building of physical facilities and on the mobilization and organization of manpower. In addition, in countries in which national extension projects were financed by the Bank, the level of in-service training was substantially increased. The Bank's participation in the financing of projects was made contingent upon national extension organizations providing, at least in part, the necessary operational resources. Thus, Bank-supported extension projects or components tended to cover all categories of capital and operating expenditures during the projects' implementation. Extension administrators in the ten countries studied were not only aware but also appreciative of this fact. This aspect of the Bank's support, although positive, raises two important questions.

First, there is evidence from some countries to indicate that the level of resources allocated to extension during the time the Bank was financially involved went beyond what the countries were able to sustain once the Bank terminated its support. Despite legal and informal assurances given by these countries to the Bank to continue funding extension services after the investment period, there was a decline in the support provided to extension after the project period. This decline was below the level required to ensure the proper operation of extension. For example, support for extension in area development projects was often not sustained after the Bank's involvement in the project ended; this was the case in Kenya, Mali, Nigeria, Sudan, and Turkey. The same problem was faced in Thailand: provinces that no longer qualified for funds borrowed from the Bank had to revert to the normal operating budget. Similar difficulties arose in Indonesia.

Second, the Bank basically supported the quantity aspect of extension manpower; the quality aspect was supposed to be addressed through in-service training. However, because extension agents have had insufficient education (up to the primary or the secondary level in most cases) and because the local capability for in-service training was limited in most countries, it appears legitimate to question whether in-service training would be adequate to improve the skills of those responsible for the knowledge transfer function, which requires as broad a set of skills as those needed to develop, modify, or adapt the technology itself.

Although the Bank's financing contributed to a better balance in the resource allocation within extension services during the investment periods of projects, it seemed that the countries' capacities to sustain those resources after the project period could have been better assessed. Moreover, as in research, the Bank's funds tended to emphasize the addition of new resources (particularly manpower) over the improvement and management of existing resources. This tendency was particularly noticeable in countries in which the Bank suppported full-scale extension projects (Brazil, Indonesia, and Thailand among them).

From a resource allocation point of view, the most viable type of support provided by the Bank in countries short of manpower and in countries facing severe budgetary constraints may turn out to be that provided in Kenya, Mali, and Nigeria.[3] In these countries, existing resources were intended to be used to transfer technology in area development projects and to balance the distribution of resources between extension activities and those of supply and services. Sudan provided another example of support being provided to the extension function (but not to the extension organization) that was designed with due consideration to the country's resource limitations: the Agricultural Bank of Sudan took the initiative in providing its clients with technology transfer services, including extension, as part of its supervised credit scheme. This initiative was successful and was adopted as a model in the agricultural services project supported by the Bank.

The lesson to be learned from these examples is that an imbalance in resource allocation does not have to be corrected by creating an imbalance elsewhere, nor does it have to be corrected overnight. Because extension was neglected in favor of supply and service functions, countries need to balance their support to various functions in such a way that these functions remain useful and

3. In Kenya, the IADP approach was based on the "farm management" approach to extension; this included farm planning and the coordination of the supply of farm inputs. The second IADP project, however, encountered major problems that resulted in its cancellation. One of the problems was the lack of coordination among agencies responsible for various services.

complementary to one another. Each country should be able to afford to provide these services to its farmers; this requires, first, careful planning and sequencing of project inputs (recognizing the country's limitations) and, second, proper management of existing resources and selective development of new resources.

Extension Methodology

At one end of the spectrum, agricultural extension is a communications task: transferring information about new farm practices from research to potential users and getting feedback from users to researchers. This is all that would be needed if the technological practices being introduced did not require major changes in the existing production process. Farmers could simply substitute the new practice for an old one and expect a higher output. For example, a crop variety could be rendered more efficient (that is, provide a higher yield) by the application of fertilizer or other inputs.

At the other end of the spectrum, agricultural extension can be seen as an education task consisting of both the communications task and the added task of helping farmers adapt their production process to take full advantage of the technological practice suggested. This role would apply whenever the technological practice being introduced led to major changes in the existing production process and required farmers to relearn how to combine their factors of production optimally.

The Bank's support for agricultural extension in national projects has favored heavy reliance on demonstrations, small group meetings, and farm visits.[4] To be properly implemented, this approach had to be labor-intensive. But because extension workers were often poorly educated, their task consisted, for the most part, in transmitting a technical message at regular periods to farmers. Therefore, the extension agent became more of a person-to-person communicator than an extension agent. This approach appears to have worked in two cases: those in which the objective was to

4. This applied particularly to India, Indonesia, and Thailand, where the training and visit technique of extension was utilized. Appendix 5 provides a further explanation of extension methodology.

assist farmers in acquiring technological elements that did not substantially alter existing production processes and those in which the emphasis in the extension agent's job was more on transfer of information than on education.

The Bank has supported communication by emphasizing a variety of media, including printed materials, displays, and radio programs. In Brazil, India, Indonesia, and Thailand, the Bank has supported both the communication and the education aspects of extension. The design of such support and its implementation were, however, basically the same in each country despite differences in the countries' farming environments and in their availability of extension resources. Thus, mass communication methods, which are primary tools of extension at the "awareness" and "interest" stages of the dissemination process, were used in these countries indiscriminately.

Moreover, the lack of skilled extension personnel often prevented the education task, which is the substance of extension at the "evaluation" and "trial" stages of dissemination, from being performed. For these reasons, the Bank's support for extension appeared to have helped in the introduction of a communication-oriented extension methodology that was suitable for the diffusion of simple technological practices. This methodology also attempted to introduce improved work habits such as discipline and accountability. The lack of skilled extension personnel hindered the development of the education aspects of extension.

In most agricultural extension projects (in India, Indonesia, and Thailand) and in some extension components (particularly in Turkey and Morocco), the Bank's support was aimed at strengthening extension and at separating it from the supply and service functions of technology transfer. The emphasis was on the regularity of visits to contact farmers by field extension workers and on the regularity of on-the-job training during which extension workers received the message to transfer to farmers. The monitoring units in several states in India were beginning to find indications of the regularity of visits. This regularity itself represented a major achievement by increasing the awareness of the extension function and introducing improved work habits and procedures.

In commodity development projects, the support to extension stressed the education aspects of technology transfer. In projects involving dairy and sericulture development in India, coffee and tea in Kenya, cotton in Mali, and horticultural crops in Morocco, an effort was made to identify proven technological practices; the extension effort was defined in relation to the nature of the technology and with good knowledge of the clients' capabilities. The cases of dairy and sericulture in India and of cotton in Mali are worth studying in detail and emulating in other places where conditions are appropriate.

In area development projects in general, and in the agricultural extension project in Brazil in particular, the Bank supported a balance in the functions of technology transfer—from an almost exclusive concentration on input supply and services to an increased emphasis on extension. In these cases, although the shift was slow (compared with full-scale extension projects in India and Thailand), it resulted in less institutional disturbance in the existing system of technology transfer. Agents supplying inputs and services did not have to be transformed into extension agents over a short period of time, nor did countries have to strain their budgets to recruit a large number of extension agents to set up a new extension organization and introduce a new extension methodology.[5] Some extension leaders in Brazil, however, expressed frustration with the slow pace of the shift.

The preceding discussion suggests that extension (communication and education) should be regarded as a *tool* of technology transfer, which can be adjusted to serve that *end* (transfer of technology) under varying *circumstances* (depending on the nature of technology and the characteristics of clients). It is the circumstances that determine the extent and the nature of adjustment (mix between communication and education) and, consequently, the nature and intensity of the extension effort required. The extension methodology should be selected to fit the circumstances

5. It should be noted that with the possible exception of Brazil, the Bank's support for extension in area development projects was provided under conditions in which research was generally very weak. Thus, even though the approach to extension support might have been the right one, its effectiveness remained limited.

under which it is to be applied. The opposite approach of attempting to change the circumstances to fit the methodology may work under certain conditions, but it cannot be viable in the long term.

India, with its vast pool of semiskilled manpower, was equipped to pursue a labor-intensive, communication-oriented extension service, particularly because the technology to be extended did not require intensive education. When manpower with even a minimum amount of skills is scarce, an alternative to the Indian approach could be a greater reliance on the communication media to extend simple types of technological practices. However, complex technological changes require that the extension function be weighted more in favor of its education function than its communication elements.[6] This requirement poses a challenge to countries in selecting technologies for extension to farmers when the capability of their extension services is limited; it does not, however, deny the need for countries to get ready for the future by progressively building an extension service that could ultimately acquire the capability of extending all types of agricultural technology regardless of their complexity and that the country would be able to afford to equip, staff, and operate in the long term.

Organization of Support for Extension

The responsibility for extension (as distinct from other functions for transferring technology) was given at times to a specialized extension agency and at other times to a multifunction technology transfer agency or a technology development agency. Each approach has advantages and disadvantages (see Appendix 7). The advantages and disadvantages depend on the means a country has

6. Examples involving complex technological changes observed in the countries visited include crop diversification in the major irrigated schemes in Sudan, mechanized farming in Western Sudan, integration of secondary crops in paddy areas in Indonesia, and integration of crop and livestock activities in Kenya, Morocco, Nigeria, Sudan, and Turkey.

to support these functions and on the degree of complexity of the technological changes that are being contemplated.

Support for technology transfer in the countries visited was provided in a variety of contexts: from simple actions aimed at the free distribution of one input to major institution-building programs involving radical organizational changes and infusion of substantial resources. In Brazil, India, Indonesia, Thailand, and, to a lesser extent, Turkey, support for extension was meant to make an immediate impact at the farm level as well as to strengthen the extension organization (comprehensive support). In Kenya, Mali, Morocco, Nigeria, and Sudan, the support for extension was mostly for making an immediate impact on agricultural production (partial support), with limited attention being paid to longer-term institution-building objectives.[7]

Among the countries that chose the comprehensive approach, India provided the closest example of compatibility between what was expected from the support to extension and the means available for the country to achieve it. India chose a communication-oriented methodology with a manpower-intensive approach to technology transfer. This approach was in conformity with the country's means because India had a large pool of rural agents who were shifted from a multipurpose community development job to a single-purpose agricultural extension assignment. In addition, because salaries in India were low by international standards, the incremental costs in the operating budget were within the country's means. Moreover, because the education system in India was relatively well established, the incremental manpower needed to implement this approach was not considered a constraint.

Two elements were left out of the country's design of extension, however. First, it was not clear whether India intended to help its extension system evolve to allow a balance between the communication and the education aspects of extension, and if so, how. The education aspect of extension is important, given India's needs to transfer the more complex technological practices that its research organizations are capable of producing. Second, although

7. Appendix 6 provides a discussion of the advantages and disadvantages of specialized and multipurpose extension agencies.

short- and long-term training were emphasized in designing projects and resources were allocated for training, the quality of extension training, especially of the long-term type, was inadequate.

Thailand represented the other extreme—it provided the closest example of incompatibility between what was expected from support to extension and the means available for the country to achieve it. The extension methodology chosen in Thailand was the same as that used in India, but Thailand did not have the same access to manpower as India did, and, according to Thai officials, it faced severe budgetary constraints to maintain and substantially increase the number of extension staff. Moreover, many Thai field extension staff expressed concern over the suitability of the extension approach in the Thai culture.[8] Perhaps the same communication-oriented extension approach used in India could have been modified for use in Thailand to place more emphasis on communication equipment and less emphasis on manpower.

In Indonesia, the support to extension, although originally based on the Indian model, has not pursued the long-term institution-building objective. Because of internal disagreements over administrative territoriality and because of the country's concern with rice production, it was decided to adopt a communication-oriented extension methodology, primarily to introduce rice farmers to well-defined technological practices. At the same time, Indonesia sought to strengthen other functions related to the transfer of technology, particularly those concerned with the supply of inputs and credit services. Whether by design or otherwise, Indonesia's support for communication-oriented extension was well organized, compared with what was observed in most other countries. Nevertheless, the long-term institutional development objectives in extension would have to be defined and addressed based on an analysis of the "rice experience." Indonesia would do well to emulate its own institution-building experience in strengthening its research system, as it seeks to build a national extension capability, to avoid the present duplication and scattering of resources among departments and agencies.

8. Similar concerns about the suitability of the extension approach to the local culture were expressed in Indonesia, Morocco, and Turkey.

The experience in Turkey was similar to that in Thailand, although it was limited to area development projects. The Seyhan irrigation project was repeatedly mentioned by officials in central ministries and in the field as an example of local support being given to extension in isolation of the existing national system and on the basis of resources beyond what the country wished to allocate to extension in the long run. Turkey, however, appeared to be learning from its own experience and was attempting to improve its extension service in the context of the Bank-assisted Corum Cankiri project.

The communication aspects of extension were differentiated to a greater extent in Brazil than in other countries. It was the only country visited where communication-oriented extension was based on the use of mass media in some areas and on manpower in others. There was, in Brazil, a clear appreciation among extension administrators of the role of education in extension; an effort was also under way to expand the country's capability to provide education-oriented extension in the future. Although some progress has been made, Brazil has not yet solved the problem of extension workers' being kept fully occupied at times by input supply and service functions (the latter include social services) at the expense of the extension function.

In Kenya, Mali, Morocco, Nigeria, and Sudan, whatever minimal support was provided to extension was directed toward short-term objectives in the supply of inputs and services. This approach was generally dictated by circumstances such as weak research institutions and limited technology to extend; it was not the result of an explicit policy decision. Many extension administrators in these countries felt that it would be more efficient to limit extension agents with low levels of training to supply and service functions than to push them into extension functions for which they were poorly prepared; this was especially so in circumstances in which there was no reliable source that produced improved technology.

Regarding extension components, the Bank's documents were unclear about which of the technology transfer functions were targeted for support. Discussions in the field indicated that the Bank provided more support for the supply of inputs and services than for extension. This support turned out to be the Bank's strength because in most circumstances there was no capability for performing extension, and no technological practices that required extension were identified. Extension agents, therefore, played a major role as input supply agents and a minor role as communication agents. In the absence of research, that role was sufficient to meet the most urgent needs.

Another outcome of the Bank's support for extension in agricultural and rural development projects was the increased awareness in member countries of the important role of technology transfer functions in the development process. This increased awareness was demonstrated by three observations: (1) of the ten countries, Brazil, India, Indonesia, and Thailand have already shifted from using Bank funds in support of extension components to using them to build up their national extension services; (2) at least two other countries—Turkey and Morocco—have had discussions about the possibility of the Bank's supporting their national systems; and (3) Kenya, Mali, and Nigeria have used the Bank's support at project level to develop a model for technology transfer to be adopted nationally.

The use of the Bank's support for extension in agricultural and rural development projects could be greatly enhanced if (1) support for extension were defined in relation to the state of research in the country; (2) the Bank's appraisal reports were more explicit in defining what is meant by, or what is expected from, extension; and (3) the Bank placed emphasis on the better use of existing resources than on the addition of new resources. As in research, resource management is the weakest link in the system of technology transfer, and for that reason the Bank should direct its support to improving it.

The basic objective of broad-based extension projects to help member countries develop an extension capability has been sound. But experience shows the need for increased emphasis on the study of the social environment, the existing administrative structure and traditions, and the extension needs of the potential clientele at the time of the project's design. These aspects are a prerequisite to the selection of a strategy to build an extension system responsive to the country's needs as they evolve over time.

As with extension components, the problem in full-scale extension projects related partly to the definition of the functions to be performed. The countries and the Bank had not completely adjusted the time frame of their projects to their objectives. Thus, the Bank continued to help member countries seek immediate production results in full-scale extension projects, as it did in extension components, but within the same short time frame. The Bank also sought to help countries reform their institutions which were the result of a long-term evolutionary process. Limitations imposed on the time frame have led to the introduction of many timesaving devices (such as on-the-job training as a substitute for basic formal training and the use of consultants to substitute for local personnel). These devices had some purpose in implementing projects primarily designed to increase production of crops. This review found no evidence, however, that the same devices could be effective in fulfilling long-term objectives, such as institution building, the realization of which would require a good understanding and appreciation of the political, cultural, social, and economic conditions in the country.

Linkages

As part of the technology development and transfer system, an extension organization has to (1) be knowledgeable of, and responsive to, the needs of its clients; (2) have a source of knowledge to rely on for answering client's questions; and (3) have established channels to supply inputs and services that complement extension. The linkages between extension and these three elements are reviewed in the remainder of this section.

EXTENSION-RESEARCH LINKAGES. Bridging the gap between research and extension has probably been the most serious institutional problem in developing an effective R&E system; this is particularly so in countries in which these two functions are carried out by different organizations. Historically, attempts at linking R&E have been through individual initiative, through formal meetings organized for research and other agricultural technical staff, or through coordinating or technical committees. India was the first among developing countries to try to establish a formal link based on the system of land grant universities in the United

States. More recently, attempts at linking extension and research were centered on that portion of the R&E continuum consisting of technology testing and demonstration. The agricultural technical specialist, who has been referred to as the subject matter specialist in the extension literature, became the key actor in establishing a bridge between extension and research.

In fact, although the functions of the SMS started in extension, some research organizations have claimed that this function should be attached to research. Indonesia's research organization, for example, has been leading efforts to operationally link itself to extension. The organization made a proposal to provide subject matter specialists with work facilities and mobility to encourage them to come to research stations and, thus, keep themselves abreast of new developments in research.

Attempts by the Bank to help address this linkage issue have varied. In some instances there was no recognition of the problem or little attempt to do anything about it. In other instances the Bank encouraged formal linkages between research and extension or attempted to establish an integrated system, either by funding agricultural R&E in the same project or in complementary projects, or by building sufficient R&E components in commodity projects to formally structure the relationship.

In area development projects in Kenya, Mali, Morocco, Nigeria, and Turkey, extension components were often designed with little concern for the quality of research or for the need to link the project to established research centers. Although some minor research components were funded in these projects, there was less emphasis on technology development than on moving supplies and services to farmers in support of existing agricultural technology. In agricultural extension projects, attempts were made to formally link research and extension. For example, in India, the extension projects helped narrow the gap at the state level, while at the district level within each state, where the location-specific technology had to be modified and tested, a serious gap still existed. In Brazil, formal exchanges at the national level were minimal, but close working relationships were starting to develop at state and local levels. In Thailand and Indonesia, the links remained informal and weak.[9] Commodity development projects, particularly those that deal with export crops, have had a longer research tradition. Therefore, the extension function in these projects was generally well related to local research or to an outside source of technology.

The Bank supported the function of the SMS as a link between research and extension. Experience in Bank-assisted projects indicates that subject matter specialists may be the most difficult group within the R&E system to train and to keep in place. First, these specialists need a strong technical background, including some research experience, to enable them to fully understand research findings and output. Second, they need to be experienced production specialists who can grow a crop and understand the problems faced by farmers. Although these specialists need this combination of valuable skills, in the past they have generally been assigned to an extension organization that more than likely had a lower status and a lower salary schedule than research. In addition, in those countries where the private sector has been expanding, agribusinesses have been hiring these specialists, sometimes to serve the same clientele as those of the extension organization, though in a much narrower context and at a much higher salary.

The Bank's recognition and support for the functions of the SMS need to be expanded to train more SMSs and to upgrade the skills of those already in their posts. The job descriptions for these specialists also require attention. For example, in Thailand and Indonesia, the specialists are overburdened—with administering extension programs and training field-level agents—to the point where they cannot perform their function as the link between research and extension. Several project leaders in Indonesia, Morocco, Thailand, and Turkey have questioned the need for large numbers of field-level extension workers and have suggested that well-trained, highly motivated, well-equipped, and mobile subject matter specialists could assume full responsibility for extension and do a much better job. This argument was based on

9. In Thailand, the national research project provided for a study of the research organization, including the role of the SMS in the research centers.

the observation that problems faced by farmers, especially the more advanced among them, were often too complex for a poorly trained field-level extension worker to analyze, dissect, comprehend, and transmit to researchers; the same worker would have difficulty understanding complex technological innovations and teaching farmers how to use them.

In summary, the institutional separation of extension from research has hindered the creation of operational links between the two functions. Although some attempts have been made at improving the situation, more work remains to be done. The way to remedy the situation depends on the institutional setup in the country concerned. The remedy could take a variety of forms, such as the sharing of physical facilities, the sharing of responsibilities for specified tasks, and dual appointment for staff. For example, the job description of an SMS could include both extension and research responsibilities, the same way a university professor has teaching and research functions.

EXTENSION-CLIENT LINKAGES. Concern over the linkages between extension services and their clients stems from the need for feedback to (1) determine the usefulness of technologies being extended to clients; and (2) ensure that extension approaches are adapted to the specific needs of various groups of farmers.

Extension organizations in the countries visited were traditionally top-down institutions; they operated on the assumption that R&E personnel knew farmers' problems and goals. Attempts at correcting this misconception were concentrated on encouraging the establishment of internal feedback channels within extension services through regular contacts between extension workers and farmers. This mechanism, although an important step in the right direction, still left farmers at the mercy of the extension agent's willingness and capability to comprehend their problems and transmit them to researchers. The extension organization in such a case lost touch with clients and took on its own objectives, which included its survival. More positive steps have been taken in some cases. In the dairy project in India, farmers' cooperatives at the village level were set up and their memberships taught to look after their own interests, including their interests in the development and transfer of technology. These farmers' organizations took an active interest in the relevance and quality of the services provided to their members.

The Bank's support for extension stressed the need for feedback from farmers. However, only a few projects in the countries visited had concrete proposals and funds for establishing formal structures through which farmers could participate in the planning and programming of R&E. In most extension projects, the Bank supported reliance on feedback through field-level extension agents, but most of the extension organizations visited lacked the capability to check on the quality of such feedback.

The example of dairy farmers in India illustrated the need to provide channels through which the less privileged groups could express their points of view and, thus, make the technology transfer function more useful. Numerous case studies have indicated that small farmers, in general, and women, in particular, had the least say on either the form or the substance of extension.

Because small farmers were more vulnerable to risk than large farmers, the former were often perceived as being more resistant to change. Extension workers were naturally drawn to those farmers who were likely to find their information useful and to appreciate their help; these were, very often, farmers with large landholdings. Such actions of extension workers had implications on the types of problems fed back to the research system for the development of future technology. Moreover, because the extension function was often tied closely to the supply of inputs and services, those farmers who had their own means or were qualified for credit gained more access to inputs and technical assistance. Therefore, if the supply of inputs and services was limited, extension agents paid more attention to farmers with more economic and political power; these generally were not the small farmers.

The Bank made a concerted effort in all of the projects it supported to facilitate access to technology by small farmers. A major part of the Bank's support for extension was directed to areas where the marginal and small farmers were the predominant groups. Access to extension services by these groups has been increasing. The extent of the benefits to farmers from such access has, however, varied, depending on the country's willingness to address such issues as land tenure, credit policies, infrastructure development, and price policy. An assessment of these factors is not possible in this review.

Concerning women, it has been known that in many societies women have had different communication networks than men. In countries in which women have been principal food producers—Indonesia, Kenya, Mali, Nigeria, and Thailand, for instance—transferring agricultural technology through men would not ensure its rapid dissemination to potential users. In fact, in some societies it could be culturally difficult for male extension agents to work with women. Therefore, if women are not fully involved in extension, they are unlikely to be consulted on their production problems and on their priorities.

Judging from the case studies, recognition of women's role in agriculture has been poor at the operational level. For example, in India's Sixth Five-Year Plan, considerable attention was given to women's role in agriculture. More agricultural extension and training resources were to have been allocated for women farmers. However, not one female agricultural extension worker was encountered during the mission to India. Even in a sericulture project in the southern state of Karnataka, where women have been largely involved with silk production and processing, there was no mention of them in the project reports, nor of the need to have women extension workers. A small pilot project in Karnataka funded by the Ford Foundation has involved female extension workers in sericulture; it appears to be quite successful. The situation is not very different in other countries, except Thailand. In Thailand, 14 percent of the field-level extension agents are women, and women are represented at all professional levels of the extension system. Among the Bank project documents reviewed, many included recommendations that facilitated access by women farmers to extension services. Such recommendations, however, needed to be translated into specific programs, particularly in training women extension staff.

LINKAGES BETWEEN INPUT SUPPLY AND EXTENSION. The effective use of improved agricultural technology by farmers depends on the availability of essential inputs and agricultural services. Ensuring that inputs and services are provided when and where needed presents considerable administrative and logistical problems. Many countries rely largely on imported fertilizers, pesticides, machinery, and seeds. Even within their own boundaries, these countries often face difficulties in transporting, storing, and marketing inputs because of long distances, poor roads, and a large number of potential customers. Even when these administrative and logistical problems are overcome, questions of access among the farming population arise. Some farmers, particularly large owner-operators, often have no difficulty in purchasing and paying for inputs and services. Others are reluctant to use inputs in optimum quantities or are unable to pay for them in cash because of perceptions of risk, lack of creditworthiness, small size of the farm, or disadvantages of tenure.

The integration of extension (or rather of communication) with input supply and services was successful in several cases. For example, outstanding progress was made in raising rice yields in Indonesia through a combined extension-credit-input approach. In parts of Kenya, corn yields increased under the hybrid maize program and other area development projects of the Kenya Seeds Corporation. Crop yields increased in Mali, Morocco, Sudan, and Turkey under similar projects and programs. In India, for some commodities such as dairying, input supply became adequate and was well related to the extension package delivered following a vertically integrated production-marketing approach (see Appendix 7).

The Bank's support for the supply of inputs and related agricultural services was closely coordinated with its limited support for extension in agricultural and rural development projects. In full-scale extension projects, however, the supply and services functions of technology transfer were separated from the extension function, and no parallel effort was made to institutionally fill the void created—except in India. This issue is particularly important in countries in which the private sector is hindered by regulations on supply or price and in which marketing channels are not well developed.

Appendix 1

Organization of National Research and Extension

A POLICY DECISION WAS MADE in Indonesia to integrate all activities involving the generation of agricultural technology in the country into the Agency for Agricultural Research and Development (AARD) attached to the Ministry of Agriculture. The link between research and policymaking at the national level is further strengthened through the agricultural research board chaired by the minister of agriculture and composed of the directors-general of the technical departments of the ministry and the heads of the research, extension, and training agencies. This board is responsible for national research policy and the allocation of funds. This strong link has allowed AARD to establish a consensus on priorities for developing an agricultural research capability in the country.

Serious attention at the policymaking level to technology transfer in Indonesia preceded that given to technology generation. The advent of the high-yielding rice varieties and the country's serious dependence on rice imports prompted the government to support a technology transfer program for rice, which covered both the supply and the service of technology and, to a lesser extent, of extension. This program was conceived and planned by the agency responsible for food crop production at the federal level and implemented by agencies of the regional governments. Other agricultural commodities did not benefit from this program, or did so only marginally. Therefore, when AARD was established, the government decided to establish the Agency for Agricultural Education, Training, and Extension to unify all extension activities in the country and to spread their benefits. Contrary to research, however, this agency did not get the policy support it needed to assume the responsibilities allocated by a presidential decree. As a result, except for rice, the technology transfer effort remains fragmented, has no national direction, and has a marginal formal link to the country's policymaking and development planning systems.

The official link between agricultural research and policymaking in Thailand is the National Research Council, an agency of the Ministry of Science, Technology, and Energy. This council, however, has responsibility only for industrial crops; this leaves other agricultural commodities without formal representation at the policymaking level. The research system in Thailand is being reorganized, but its emphasis on the operational aspects of the system remains. The issue of linkage between the research organization and the policymaking level has not been addressed yet, nor has the policy decision of entrusting national responsibility for agricultural research to an agency been unequivocally made.

Technology transfer activities in Thailand, outside of industrial crops, developed initially around rice. These activities progressively expanded in response to various needs identified independently by various departments. Initially, the technology transfer concentrated on supply and service activities. In 1977, the government established the Department of Agricultural Extension with responsibility for extension in areas other than those under the jurisdiction of the departments for livestock, fisheries, land development and cooperative promotion, the tobacco monopoly, and the rubber replanting aid fund.

Research and extension in Sudan developed around irrigation schemes centered on cotton production. Repeated attempts have been made to expand the scope of these services to include other agricultural commodities and the rainfed areas, to conform with the country's development needs. A national research council was established and given responsibility for planning and coordinating national research, including national agricultural research. This council was not, however, given the means to assume that responsibility. As a result, decisions concerning agricultural research were left to agencies in charge of implementing research activities. The outcome has been a very limited input into research from the policymaking and development planning levels of government and a lack of consistency between the direction of the country's development efforts and that of its research activities.

Technology transfer activities in Sudan have been almost

exclusively limited to supply and service activities (mainly in the irrigation schemes controlled by public corporations). An attempt at introducing extension activities along with the supply and service of technology has been successfully made by the Agricultural Bank of Sudan, which has begun to supervise credit schemes in both irrigated and rainfed agriculture areas. The organization of technology transfer, however, is still being debated in the country.

Appendix 2

Impact of the Bank's Support on the Development, Allocation, and Management of Resources

THE TEN COUNTRIES included in this study can be separated into two groups. In the first group—comprising Brazil, India, Indonesia, Sudan, and Thailand—a favorable assessment may be justified on the grounds that the Bank's lending has caused a major increase in the resources available to research, not only because of external funds, but also because of policymakers' commitment to increasing funds allocated for research from local sources. In these five countries, the Bank's support also contributed to the geographical expansion of the research network and to a better coverage of crops, particularly those produced by small-scale farmers. Such crops were historically neglected, particularly in Brazil, India, Indonesia, and Sudan. In Sudan and Thailand, the Bank's support had been questioned on the grounds that the countries were unlikely to sustain the level of support to research once the Bank's funds stopped.

In Thailand, policymakers' reluctance to define the likely level of long-term support for research has made research administrators reluctant to integrate into their normal operations what they perceive to be "a temporary injection of outside funds into their research system." The administrators instead have suggested that outside funds be used to undertake research activities parallel to their normal work. Thus, the decision on integrating Bank-supported activities (or facilities) into the research system would be left open until the time when policymakers are faced with the problem of substituting local funds for exhausted outside funds. The inability of the country to select proper institutional changes and adopt needed policies to support the introduction of external funds has caused delays in the implementation of the project and may cause changes in its objectives.

In Sudan, the Bank requested at the outset that the research project be implemented by an entity independent of the national research organization. This request eliminated the disagreement (which happened in Thailand) between policymakers and research administrators; however, it did not eliminate the discontent of the latter group with the arrangement selected. Research administrators in Sudan feel that better use could have been made of project funds if they were integrated into the budget of the national research organization and if their level was adjusted to what policymakers would accept as a reasonable level—one that the country could sustain in relation to its other commitments. Unlike in Thailand, there were no indications in Sudan that research administrators have shared their views with policymakers in the country or with the Bank's staff.

In the second group—comprising Kenya, Mali, Morocco, Nigeria, Sudan, and Turkey (Sudan was also included in the previous group)—the Bank has provided limited support to research, often in the form of minor components in agricultural and rural development projects. Even though the Bank's financial input has been minor, administrators of national research organizations have taken a positive view of its limited impact on resource allocation within research for two reasons. First, the Bank's input helped draw the attention of national research organizations to commodities that had not been recognized, including those benefiting traditionally neglected areas or groups of people. This was the case in Turkey (forestry and livestock), Morocco (rainfed agriculture), Sudan (crops and livestock in the western and southern regions), Kenya (natural pasture), and Nigeria (livestock). Second, support to research through components in agricultural development projects has demonstrated to researchers the close relationship their work has to the country's overall agricultural development effort.

While recognizing these positive aspects, most administrators of national research organizations tended to agree that allocation of resources for research in a specific project, while justified in the past, should be avoided in the future. At the very least, these resources should be made part of an overall program to support the national research organization on a sustainable basis. In the administrators' view, research components, which in the past were a tool for getting researchers and policymakers to regard research as a means for development, should be transformed into a tool to enhance the productivity of the national research organization and its contribution to the development process.

Some countries—Mali, Morocco, Sudan, and Turkey among them—have viewed the strong emphasis on the use of consultancy services in research components as unwarranted. On-the-job-training aspects of consultancy services have also been questioned on the grounds that consultants, who may be qualified to undertake research, are often not qualified to do the training, or that local trainees are not suitable for receiving training. Experience shows that consultants were useful when the research component was well defined and where their role in implementing it was well specified.

Appendix 3

Research Organization

IN INDIA, THE NATIONAL RESEARCH POLICY is formulated by the Indian Council of Agricultural Research (ICAR), a semiautonomous organization within the Ministry of Agriculture. The chairman of ICAR is the minister of agriculture. ICAR has a substantial staff and supporting budget. The council channels funds to its thirty-four central institutes and national research support bureaus and, through the fifty nationally coordinated research programs and ad hoc projects, to the twenty-two state agricultural universities responsible for decentralized research. Both the ICAR institutes and the state universities have regional stations, substations in each ecological zone, and adaptive research projects.

Brazil has an autonomous national research corporation (EMBRAPA). Its board of directors puts together a national research program as the basic guideline for research policy. Unlike India, Brazil has no broad-based agricultural research council, but EMBRAPA has a strong support staff for planning and monitoring research and has good links to national planners. In addition to twelve commodity research centers and four service units, EMBRAPA also plays an important role in providing funds and helping plan research at seventeen state-level stations located mainly in the poorer and less-developed states; it also operates three special ecological zone stations of its own. Three developed southern states in Brazil operate sizable research services of their own, outside the purview of EMBRAPA.

Indonesia has an umbrella organization, the Agency for Agricultural Research and Development (AARD), which functions under the Agricultural Research Board chaired by the minister of agriculture. Its budget has to be approved by the minister of state for research and by the National Planning Agency. AARD has a two-tier structure, with five central research institutes representing major subsectors of agriculture, each managing several operational institutes throughout the country. The heads of the central institutes report directly to the head of AARD, as do the heads of six other centers providing specialized support services to research. The support services include programming and monitoring, agroeconomic research and statistics, and data processing. These functions more or less parallel those performed by the national bureaus of ICAR in India and the four national support units of EMBRAPA in Brazil.

Among the other countries studied, Mali has a similar structure; its central research coordinating agency (IER) has service units and a national committee. Mali, however, has not yet achieved a consistent long-term government commitment to research, which would provide sustained operational funding under the management control of a given agency. This situation prevails despite the fact that Mali allocates 0.8 percent of its agricultural GDP to research.

Turkey is the only other country that seems to have developed a system permitting independent control of budget allocations for research operations and investment: researchers have to submit their programs for discussion and approval by the State Planning Organization (SPO) attached to the prime minister's office. The SPO does not seem to play much of a role in the coordination of research planning, however, because under the Turkish system research is fragmented among three ministries, eight directorates, sixty experiment stations, and several universities.

The other five countries have no effective central organization. Nigeria has had several agricultural and scientific research councils since 1971. The country still lacks a satisfactory professional mechanism for setting overall priorities or for coordinating the programs of its national institutes. Kenya, which allocates more than 1 percent of its agricultural GDP to research, has a highly fragmented research structure involving eight ministries or agencies but no functional central coordination. The same applies to Sudan and Thailand, although they have a theoretical capability through a national research council or equivalent body. Morocco, Sudan, and Thailand are reviewing their research structures.

Sudan has a semiautonomous Agricultural Research Corporation (ARC) under the Ministry of Agriculture, Food, and Natural Resources. Numerous subsectors, however, including irrigation, livestock, and university research, are outside its control. A similar degree of fragmentation of funding and operational responsibility exists in Thailand, where advice to the prime minister and coordination of research policy, including agricultural research, are the responsibilities of the Ministry of Science, Technology, and Energy. Eight ministries or departments and nine directorates are involved in centralized research; only a limited number of agricultural research activities are decentralized, partly because of inadequate funds, but also because of the reluctance of qualified scientists to live in remote areas.

Although Morocco has four directorates or agencies involved in research, seventeen central stations, and fifty regional or other stations, it has no council or other overall entity for the planning and management of research. Morocco is considering a major consolidation of its programs into a new National Institute of Agricultural Research.

Appendix 4
Research Links

IN 1980 BRAZIL HAD collaborative programs with forty-eight foreign institutions, including six international research centers, and six other national programs. EMBRAPA, the national research corporation, drew on these centers for germ plasm, technical training, consultancies, and technical information.

India has good linkages with several research institutions in other countries and with the international centers. These linkages are established centrally with the Indian Council of Agricultural Research and between individual ICAR institutes and the international centers. At the state level, the universities can obtain genetic material from international centers, or elsewhere through the medium of the Genetic Resources Bureau of ICAR, but apparently not directly. Although many institutes collaborate with the International Crops Research Institute for the Semi-Arid Tropics (ICRISAT) and the International Rice Research Institute (IRRI), their external linkages, whether for research, information, or training, are restricted by the intermediary role of ICAR.

Indonesia has developed very strong linkages with IRRI, which has had a team of scientists working in the country since 1972 to help with breeding, station development, and training. The International Center for Maize and Wheat Improvement (CIMMYT), the International Potato Center (CIP), and ICRISAT are also collaborating with Indonesia's Agency for Agricultural Research and Development. Australia, Japan, the Netherlands, and the United States are among the countries that have bilateral research linkages with AARD.

Thailand has had a long and fruitful collaboration arrangement on rice with IRRI scientists who work in the country with Thai scientists; it also has a strong training link. Thailand works with the Rockefeller Foundation, CIMMYT, and ICRISAT on corn and sorghum breeding and training. In addition, Thailand has research linkages with the research institutes in Australia, with the Ford

Foundation, and with the United States Agency for International Development (USAID).

Kenya works closely with the International Institute of Tropical Agriculture (IITA), the International Livestock Centre for Africa (ILCA), and ICRISAT on commodity programs to develop low-cost technology for low-rainfall areas. It has a program with CIMMYT on cereal breeding, economics, and farming systems. The International Laboratory for Research on Animal Diseases, based in Nairobi, is involved in work on trypanosomiasis and East Coast Fever. Kenya also has links with research agencies in Canada, the Federal Republic of Germany, the Netherlands, the United Kingdom, and the United States. Most of the bilateral agreements provide funds as well as technical assistance.

Mali works with ICRISAT and with the West Africa Rice Development Association (WARDA). Bilaterally, its main research cooperation is with France, Germany, Japan, the Netherlands, and the United States.

Although Nigeria has IITA in Ibadan, there are indications that the national institutions are not making effective use of this resource, nor of ICRISAT, ILCA, ILRAD, or CIMMYT. It is, perhaps, only a matter of time before they begin to use the resources of the international centers. IITA's relations with Nigerian research agencies are improving, although the center obviously cannot devote all its efforts to work in Nigeria alone. ILCA and ICRISAT are both developing cooperation with Nigerian institutions. Bilateral agencies, including those from Germany, the United Kingdom, and the United States, are also collaborating with Nigeria's research agencies.

Morocco's linkages with outside research institutions are weak, though with France, and more recently with American institutions, they have been strong. Linkages with the international centers are confined to limited contacts with CIMMYT and the International Center for Agricultural Research in the Dry Areas (ICARDA); this may reflect the weaknesses of its own staff, which hamper the country's ability to work with outside centers.

In Sudan, ARC has been cooperating with the international centers ICARDA and ICRISAT. There are also strong bilateral training linkages with the university system of the United Kingdom. New program development for the western Sudan, with the support of the Bank and USAID, has opened up channels of communication with a consortium of universities in the United States.

Turkey has established strong links with several international centers, especially CIMMYT, which has participated with USAID and the Rockefeller Foundation in a successful collaborative program with Turkish scientists to raise yields of wheat, barley, and legumes. Cooperation in research is also under way with CIP, ICARDA, and ICRISAT, as well as with Germany and Sweden.

Appendix 5
Extension Methodology

AGRICULTURAL COMMUNICATORS DEFINE technology transfer as the communications task of transferring information about new technology from the researcher to the farmer and of transmitting farmers' problems back to the researcher. For agricultural extensionists, although a two-way communication is essential, technology transfer also involves helping clients learn how to use new technology to achieve their goals. In short, an agricultural communicator would emphasize information transfer, while an agricultural extensionist would emphasize educating clients to use technology effectively. Agricultural communicators tend to rely more on mass communications, while agricultural extensionists tend to work more with individuals and groups. Agricultural communication makes intensive use of capital while agricultural extension uses more skilled manpower.

The mix of communication and extension needed in a country depends on the nature of the technology being extended, on the farmers' level of technological sophistication, and on the existence of alternative sources of information, including the private sector and farmers' organizations. Agricultural technology can be simple or complex. For example, a new high-yielding rice variety can be a simple form of technology if the farmer merely has to substitute one variety for another. Transferring simple technology to farms may be largely a communication task. The farmer does not have to learn new practices or techniques. However, if the farmer needed to learn a new production process, then technology transfer would become a more complex task. A good example would be the discovery of a new rice variety that matures in 90 instead of 130 days; this would allow the farmer to grow two crops a year instead of one, but to get the second crop the farmer would have to direct-seed rather than transplant the first crop. In addition, the farmer might have to change the timing and the rate of fertilizer application for both crops and apply an insecticide on the second crop to control insects. Clearly, transferring more complex forms of technology requires more educational input. In the example above, there is a need for both communication and extension, particularly extension.

Moreover, the approach to the transfer of any technology varies with the cultural and socioeconomic characteristics of the potential clients. For example, farmers in a relatively closed environment tend to rely on more personalized forms of communication, especially with people in whom they have developed confidence. As their environment opens up, their medium of communication becomes less personalized, and they acquire skills to seek and process their own information. Thus, any extension methodology has to recognize the nature of the client's decisionmaking process and be flexible enough to adjust to changes over time.

Appendix 6

Organization of Support for Extension

A SPECIALIZED EXTENSION AGENCY has generally been advocated for three reasons: (1) the staff members would have the opportunity to concentrate on extension, including its communication and education components; (2) the staff would be relatively homogeneous, which would enable the organization to monitor and evaluate their performance more effectively; and (3) there would be a high probability that once resources are allocated to extension, they would be used for that purpose.

There are three disadvantages to setting up a specialized extension agency, particularly in those developing countries in which institutions are very weak. First, a specialized agency would require the parallel development of other agencies to perform other technology transfer functions. Second, it would create a need for coordination among agencies responsible for technology transfer, on the one hand, and between these agencies and policymakers and planners, on the other. Third, it would increase the competition for scarce manpower resources and funds when resources are insufficient to meet the existing needs of all agencies.

An agency made responsible for extension, input supply, and services has major advantages under special conditions. For example, if an area or a category of farmers needs special development assistance, some or all of the technology transfer activities can be combined in one organization and the size of the staff can be tailored to meet the exact nature of the functions to be performed. The disadvantage of this approach lies in the risk that the relatively less demanding but more visible functions (such as the supply of inputs and services) will distract the attention of the management and the staff from the equally important function of extension, where and when it is required.

The selection of a form of organization for supporting the technology transfer functions in general, and extension in particular, has to start with a clear definition of the objective and the means available to achieve it. This may sound obvious, but it was a missing element in all the systems reviewed. The objective needs to be specified in relation to time. Thus, a short-term objective (for example, the transfer of a defined technological practice) should be clear—not only the substance, but also how it relates to a longer-term objective (institution building). The two are not mutually exclusive, but they may require different policy decisions and strategies. Of the means available for seeking these objectives, a distinction should be made between wanting to make the best possible use of inadequate resources to perform a function and the building of a capability to achieve perfection in the performance of that same function in the future. Again, the two alternatives are not mutually exclusive. The first may precede, and even lead into, the second.

Appendix 7

Extension–Input Supply Links

THE RANGE AND THE COMPLEXITY OF PROBLEMS that have to be faced in providing inputs and services have led governments to try numerous policies and measures to overcome them. The policies include subsidies of inputs, transport cost and interest rates, guaranteed prices, tenure or land reforms, integrated development projects with emphasis on small farmers, and support for farmers' associations and cooperatives.

At present, the use of these arrangements varies widely both among and within countries. In some countries, special production-oriented programs have been set up to deliver a package of inputs, credit, and advice to farmers. Sometimes these programs are commodity-based, sometimes area-based, though often with commodity biases. For example, in Mali, under the "operations" projects, a mixed commodity–area development approach is being followed. Extension personnel are made responsible for delivering an integrated package of credit and subsidized inputs for cotton and other major food crops (millet, sorghum, groundnuts, and rice, for instance).

In Nigeria, the newly formulated area development projects focus on the coordination of extension services with the supply of agricultural inputs, support services, and infrastructure development. These services and inputs are to be provided through farmer service centers and seed production farms. Credit, however, has been and continues to be a bottleneck in Nigeria. In Brazil, the input delivery system in the three richer southern states is largely in the hands of the private sector and the cooperatives, and it works well. In other parts of the country, particularly in the northeast, there is a greater degree of government intervention to try to improve access to inputs by small farmers in resource-poor areas. In Thailand, where the private sector has historically played an important role in supplying seed and fertilizers, the demand for inputs has been restricted because of poor input-output price ratios.

In India, different states have developed varying means of input supply to farmers. Cooperatives have played a major role in many states, and some staff have been assigned full-time responsibility for nonextension tasks.

Appendix 8

What Some People in the Field Had to Say

Indonesia

Some officials were of the opinion that projects were supervised too frequently and too superficially; they felt that two or three people visiting a project for ten days could not supervise a national system effectively. Supervision is preferred once a year, with a larger team visiting a project for an extended period and with some continuity of staff. Unfortunately, the Bank's resident mission in Jakarta is too burdened to provide adequate technical support. The government is trying to change extension service too fast; this has resulted in the quality of personnel being sacrificed.

Because Indonesia's research system is under development, officials have recommended that the country start with research components based on identified needs—a temporary solution that might lead to the strengthening of national research capability. There has been limited participation by local agencies in the design of projects. For technical assistance, consultants have often been selected on the basis of their technical capability to play a lead role in institution development. Indonesia has been reluctant to pay for technical assistance because of its belief that local staff were not capable of supervising technical assistance and using it to the country's advantage. The use of technical assistance through cooperation between international research centers and national research programs has been more beneficial than the direct use of technical assistance in national programs. More reliance on the

Note: These comments from officials in five countries are reported to highlight some perceptions people have in the field. These are notes taken in the field by a mission member who visited the five countries listed. No deliberate effort was made to collect views in the other countries. Although this review does not concur with some of the comments, they are, nevertheless, reported because it is important for the Bank to be aware of the perceptions of its developing member countries, regardless of whether these views are right or wrong.

latter system is required. The country should provide logistical support to the visiting specialist, while his or her salary can be paid for by the cooperating agency. This arrangement would ensure that the technical assistance is properly supervised and is relevant to research work.

Many officials feel the Bank should have stressed institution building in extension. If the Bank is going to get involved in institution building, it will have to make changes in its monitoring process to permit adjustments in project design during implementation; this may cause delays in implementation. Institutional issues have to be handled in a national context. An institution should not be strengthened at the expense of another. Creating new agencies is not an efficient way of addressing the issue of why existing agencies do not work. In the extension methodology being supported by the government, communication takes place from extension to farmers in one direction: "I am telling you, you listen to me." This should be changed to a two-way system: "I can learn from you and you can learn from me."

Many senior extension staff have indicated that village extension workers are not a good medium for providing feedback; they are tied to "work targets" and deal mostly with progressive farmers. Small farmers need to organize themselves and use their organizations as a medium for generating feedback. Extension should not be oriented toward production only; it should serve the rural community too, particularly in the fields of nutrition and technology consistent with the goals of (1) adequate employment in rural areas; (2) a healthy environment; and (3) family labor-sharing arrangements. Long-term planning is also needed to synchronize support for the development of manpower, physical facilities, and the programming and implementation of research and extension activities.

Several planning officials feel that R&E organizations tend to react to the availability of funds rather than plan their work programs and justify the use of funds. Officials from the Ministry of Agriculture appear more concerned with the mechanics of extension than with the purpose of extension. Thus, in the process of pushing an extension methodology, they become inflexible and lose track of the purpose for supporting extension in the first place.

Thailand

Senior staff in the field have indicated that extension places too much emphasis on a strict schedule of farm visits, even though technical information offered to farmers is limited; they have also indicated that extension workers are precluded from providing any other information.

Officials in the Ministry of Agriculture feel that coordination of research and extension by means of technical committees does not work because committees do not have the authority or the means to enforce their recommendations. The nature of extension and its intensity should be related to farmers' needs. In farm areas that require intensive cultivation, farmers often need nontechnical information (on marketing, prices, input supply, and credit) more urgently than technical information. In other areas, farmers could benefit greatly from technical know-how already available in the country and utilized by more advanced farmers.

Senior staff have indicated that the Bank's supervision input is very important in Thailand because it helps accelerate procedures and provides support to the implementing agency in solving problems. For this reason, there is a need for flexibility in the timing and frequency of supervision missions. The Bank's support is fragmented among uncoordinated departments. Moreover, Bank-supported activities are not complementary; they lack synchronization in timing (research, for example, lags behind extension). There is also no coordination of the activities supported by different aid agencies. These agencies have supported different methodologies in the same field, and each agency has required its own staff to work at the provincial level. The result has been a continuous reshuffling of Thai manpower to meet loan requirements or grant conditions that are based on limited appreciation of the country's needs and culture. One way to link research and extension would be to give the subject matter specialist a dual official appointment; the content of the job could be part research and part extension.

Staff in the field felt that the role of the village extension agent would have more meaning, under Thai conditions, if it did not preclude provision of other urgently needed services, such as credit, input supply, cooperative organization, marketing, and so forth.

Extension workers should be part of the community, not just messengers who provide technical information. The government is encouraging the private sector to get involved in the provision of services in "progressive areas." In such areas, public services should not duplicate those services; they should, instead, be channeled to less advantaged areas that, for the time being, are less attractive to the private sector.

Field staff have also indicated that extension methodology should not require drastic changes in people's behavior overnight. There is always a reason why people behave the way they do; it is not ignorance or stupidity. Both the government and donor agencies should try to understand why people behave in a particular manner before asking them to conduct themselves differently. Village-level workers are not qualified to identify problems and analyze them. Thus, even if they are asked to deliver a simple message, they cannot be relied upon to provide meaningful feedback. The subject matter specialist could assume this function if his or her job were altered to allow more time in the field and at research stations, provided the person was given the means to be mobile.

Moreover, field technicians have viewed the role of consultants as a means for ensuring that the project is implemented according to design. Thus, Thai field staff have no opportunity to adapt proposals made in project documents to local conditions and to local needs. Regularity of visits by extension workers has become the objective of extension. Nobody seems to be asking whether any extension work is taking place at all.

Sudan

Both planning and research staff believe that the time frame of the project allows for, at best, the completion of the construction of facilities and the procurement of major equipment. A limited amount of time is permitted for the actual research work and for getting national organizations to use the new facilities and equipment.

Senior staff in the National Research Council indicated that at present the programming of research is left to the discretion of individual researchers. No organization is responsible for formulating research policy and for relating research to development goals. The country needs to allocate the available research manpower to do research on existing agricultural practices, identify better practices, and package these practices for delivery to farmers who are not using them.

Many officials agreed that the country needs a multipurpose extension service that would concentrate, first, on communicating known improved practices and generalizing their use and, second, on helping farmers gain access to inputs and services. The emphasis of outside support should be on strengthening existing institutions by working with them. If an existing institution is not allowed to take part in implementing a project while outside financing is involved, that institution cannot be expected to perform miracles once it takes over the activities of the project when external funds are no longer available and internal support is uncertain.

Officials in several agencies indicated that the delays in project implementation were caused by (1) inadequate design of a project (for example, realities of the project area, such as the availability of housing or manpower, political constraints, and so on, may not be reflected in the design); (2) lack of coordination among financing agencies at the time of project design; (3) limited incentives for research scientists to leave headquarters and go where they are needed; and (4) problems posed by outside technical assistance.

Technical assistance problems can arise, first, in the identification of qualified people. People who apply for jobs may be starting their career and therefore lacking in experience, or they may be about to retire and thus too old to be productive in the project's harsh conditions. A second problem arises from the amount of control aid agencies have over the employment of technical assistance staff because these agencies have de facto control over the use of funds.

Senior staff in the Gezira and Rahad schemes indicate that for agricultural production corporations and agricultural development schemes the present system of funding research is not adequate. The inadequacies arise out of (1) uncertainty in the flow of funds from the public budget, and (2) a decrease in the normal budgetary allocation when the corporation or the scheme takes it upon itself to supplement the limited budgetary allocation.

Although contracting out research activities to the national research organization could be one way for corporations and development schemes to get their research needs met, such contracting could produce results only if it is kept outside the normal budgetary allocations. One way to serve the corporations and other development schemes would be for the government to pay the cost of research infrastructure and basic salaries from its budget and require corporations to pay for research activities.

Planning officials felt that the Bank's support for research and extension is very limited (when compared with its support for production-oriented activities) and very fragmented (very small amounts are attached to various projects). Moreover, research and extension components often end up being ignored, wittingly or unwittingly, during project implementation. Outside funding agencies are always eager to have their "own project," and too often they make it a condition for providing the funds. This condition has led to the "ruins of development" commonly encountered in Sudan, whether in the form of ghost buildings and unused equipment or in the form of functionless agencies. Outside funding agencies must recognize the vulnerability of countries like Sudan and help them build strong institutions that would ultimately bear the stamp of Sudan even without requiring that, in the interim, they bear the stamp of the funding agency.

Turkey

Staff of the Ministry of Agriculture indicated that under present conditions it was not efficient to provide extension support by itself. Technical advice may help, but if given alone it would take a long time for it to become useful and it could benefit only some farmers. A combination of proven technical advice is needed along with the supply of credit and inputs. (A lot of technical advice is available in the country.)

Field extension staff were of the opinion that the problem with the "old extension approach" was not the multiplicity of functions, but rather the lack of organization, the absence of a defined purpose, and the limited means for work. Thus, changing to a "new extension approach" may serve to hide the problems for a while, but it would not solve them.

Senior research scientists indicated that a major problem in research lay in the fact that research scientists had to handle administrative responsibilities; this was the only way for them to achieve professional growth.

Morocco

Senior staff in several development schemes indicated that the extension job in the irrigated areas required qualified personnel who, on the one hand, would be capable of educating farmers in the use of complex technological practices and, on the other hand, would be capable of identifying problems on the basis of observed symptoms and transmitting these problems to research. Such extension personnel can be limited in number, but they need to be mobile and motivated to serve farmers. The extension job in the rainfed areas, at this stage of development of the agricultural sector, should concentrate on spreading information among farmers about practices followed by the best among them and about programs aimed at helping them. This responsibility should be that of a few well-trained people, with the necessary mobility and the communication equipment, who could use the information generated within the country.

Development projects cannot rely on the national research organization for their research needs because they are designed to be production oriented, while the national system still requires long-term institutional support to make it productive.